Technology Transfer in Consortia and Strategic Alliances

International Series on Technical Innovation and Entrepreneurship

General Editors: George Kozmetsky and Raymond W. Smilor

Technology Companies and Global Markets: Programs, Policies, and Strategies to Accelerate Innovation and Entrepreneurship (1991)
edited by David V. Gibson

University Spin-Off Companies: Economic Development, Faculty Entrepreneurs, and Technology Transfer (1991)
edited by Alistair Brett, David V. Gibson, and Raymond W. Smilor

Technology Transfer in Consortia and Strategic Alliances (1992)
edited by David V. Gibson and Raymond W. Smilor

The Technopolis Phenomenon: Smart Cities, Fast Systems, and Global Networks (1992)
edited by David V. Gibson, George Kozmetsky, and Raymond W. Smilor

Technology Transfer in Consortia and Strategic Alliances

Edited by
David V. Gibson
Raymond W. Smilor

Rowman & Littlefield Publishers, Inc.

ROWMAN & LITTLEFIELD PUBLISHERS, INC.

Published in the United States of America
by Rowman & Littlefield Publishers, Inc.
4720 Boston Way, Lanham, Maryland 20706

Copyright © 1992 by Rowman & Littlefield Publishers, Inc.

All rights reserved. No part of this publication may
be reproduced, stored in a retrieval system, or transmitted
in any form or by any means, electronic, mechanical,
photocopying, recording, or otherwise, without the prior
permission of the publisher.

British Cataloging in Publication Information Available

Library of Congress Cataloging-in-Publication Data

Technology transfer in consortia and strategic alliances /
edited by David V. Gibson, Raymond W. Smilor.
p. cm.
"Based on the conference 'Consortia and Technology Transfer'
which was held at the IC^2 Institute at the University of Texas at
Austin in December 1989"—Acknowledgments p.
"International series on technical innovation and
entrepreneurship"—Jacket.
Includes bibliographical references.
1. Technology transfer—International cooperation.
I. Gibson, David V.   II. Smilor, Raymond W.
III. IC^2 Institute.   IV. International series on technical
innovation and entrepreneurship.
T174.3.T3868   1992
338.9'26—dc20   91-32415 CIP

ISBN 0-8476-7717-6 (alk. paper)

Printed in the United States of America

 The paper used in this publication meets the minimum requirements of
American National Standard for Information Sciences—Permanence of
Paper for Printed Library Materials, ANSI Z39.48-1984.

Contents

List of Figures and Tables		ix
Preface		xiii
Acknowledgments		xv
Part I	Leadership Perspectives on Technology Transfer	
1	The Importance of Technology Transfer and Commercialization to U. S. Competitiveness *Mark S. Lieberman*	1
2	Encouraging Innovation and Excellence in Technology Transfer *Grant A. Dove*	9
3	SEMATECH: Necessary but Not Sufficient *Robert Noyce*	17
4	It's Not Over until It's Transferred! *Edward A. Miller*	21
Part II	Organizational Culture and Technology Transfer	
5	Technology Transfer in Knowledge-Intensive Organizations *Bill Curtis*	27

6	Science as a Competitive Weapon: Utilizing Alliances in Global Markets *Debra M. Amidon Rogers*	39
7	Technology Transfer in a Diverse Corporate Environment *Wendell M. Fields*	57
8	Creating and Managing Consortia Cultures through Transitional Episodes *Michael S. Rubin*	67
Part III	**Policies and Procedures for Technology Transfer**	
9	Transferring Technology in R&D Consortia: Effective Forms of Interorganizational Relations *William M. Evan* *Paul Olk*	91
10	A Consortium-Based Model of Technology Transfer *Larry Novak*	103
11	Technology Transfer at the Software Productivity Consortium: A Partnership Approach *Claude DelFosse*	111
Part IV	**Overcoming Barriers to Technology Transfer**	
12	Building a Technology Transfer Infrastructure *Raymond W. Smilor* *David V. Gibson*	129
13	Technology Transfer from Academia: Prescription for Success and Failure *John F. Hesselberth*	151
14	Partnerships with Industry and the Oak Ridge National Laboratory *Warren D. Siemens*	157

Contents • vii

Part V	Perspectives on Japanese Consortia and Technology Transfer	
15	Outline of the Fifth-Generation Project and ICOT Activities *Takashi Kurozumi*	173
16	Comparison of Several Types of Consortia for Software Research and Development *Fumihiko Kamijo*	193
17	ERATO: A Multidisciplinary Stage for Young Researchers from Industry, Government, Universities, and Abroad *Alan Engel*	201
Part VI	Perspectives on European Consortia and Technology Transfer	
18	The European Computing-Industry Research Centre: Challenges and Answers *Hervé Gallaire*	213
19	AI Research and Its Opportunities for Technology Transfer at Siemens *Dieter Schütt*	223
Part VII	New Initiatives in Technology Transfer	
20	The American Technology Initiative: Competitiveness through R&D Joint Ventures *Syed A. Shariq*	233
21	Technology Transfer in a Tripartite Consortium *J. Grant Brewen*	241
22	A Western Technology Transfer Network for Small Business *Louis D. Higgs*	247
Appendix I	List of U.S. Consortia Registered under the National Cooperative Research Act of 1984	255
Appendix II	Consortia Filings and Membership	269
About the Editors		271
About the Contributors		273
About the Sponsors		277

List of Figures and Tables

Figures

5.1	Productivity factors in TRW	28
5.2	Individual differences in NASA software	29
5.3	Team communication by artifact	31
5.4	A programmer's communication nets	32
5.5	Variance in productivity as a function of technology use	34
5.6	The layered behavioral model of software development	35
5.7	Sources of productivity growth in different industries	36
5.8	Technology transfer lags in software engineering	37
6.1	Toward paradigm integration	40
6.2	The global environment	44
6.3	Balance of economic perspectives	46
6.4	Thesis framework	47
6.5	The Horwitch model of joint ventures	48
6.6	First tier of the lotus flower at the intraorganizational or microeconomic level	49
6.7	Second tier of the lotus flower at the interorganizational level	50
6.8	Third tier of the lotus flower at the transnational level	51
6.9	The core of the lotus flower: the customer	52
6.10	New strategy elements for global innovation management	54
7.1	Pyramid of individual development	59
7.2	Hewlett-Packard's multifaceted approach to technology transfer	60
7.3	Hewlett-Packard's expert forums to assist in technology transfer	62
8.1	The metaorganization as counterstructural transformation	70

8.2	Episodic transitions and collaborative competence	75
8.3	Transitional stages	78
8.4	Collaborative strategy, structure, and process	79
9.1	Mean satisfaction with technology transfer mechanisms	97
9.2	Typology of satisfaction with technology transfer	98
10.1	SEMATECH as a part of a larger set of entities needed to improve the U.S. semiconductor industry	104
10.2	SEMATECH's technical communications organization	105
10.3	The governance of SEMATECH	106
10.4	Member-company technology transfer from SEMATECH	107
10.5	Most effective means of technology transfer at SEMATECH	108
10.6	The types of meetings at SEMATECH and the participants	109
10.7	SEMATECH's customer window	109
11.1	A partnership of industry leaders	112
11.2	A common set of transfer mechanisms supports the technology partnership	114
12.1	U.S. R&D consortia by industry type	130
12.2	Effectiveness of MCC's technology transfer (mean scores)	133
12.3	Effectiveness of methods for technology transfer (mean scores)	135
12.4	Importance of factors facilitating technology transfer (mean scores)	136
12.5	Importance of barriers to technology transfer (mean scores)	137
12.6	Ways to improve technology transfer	138
12.7	Technology transfer variables	139
12.8	Technology transfer grid	142
12.9	Technology transfer infrastructure	144
15.1	R&D objectives for fifth-generation computers to meet the needs of the coming information society of the 1990s	174
15.2	What is fifth-generation computer systems (FGCS)?	175
15.3	Framework of FGCS	176
15.4	Organization of the Fifth-Generation Computer Committee	177
15.5	Basic configuration image of the fifth-generation computer system	179
15.6	Concept diagram showing how research and development are to progress	180
15.7	Organization of the FGCS project	181
15.8	Language system of the fifth-generation computer	185
15.9	R&D steps of system software, kernel languages, and hardware	186
16.1	Open laboratory	194

16.2	Consortium company	197
16.3	Independent division	198
17.1	Nikon's sales of cameras and semiconductor equipment	203
19.1	Siemen's corporate R&D, information and knowledge processing: scientific connections	224
19.2	Technical risk, complexity of software, and technology transfer	225
21.1	Equity members in the Biotechnology Research and Development Corporation (BRDC)	242

Tables

9.1	Percentage of consortia using these mechanisms to transfer technology	96
9.2	Relationship of transfer mechanisms to satisfaction typology	99
9.3	Relationship between technology transfer satisfaction type and performance	100
11.1	Consortium transfer matrix	117
12.1	Companies involved in six or more consortia	131
13.1	Du Pont's experience with research contracts and consortia	153
14.1	Companies participating in the Thermomechanical Model Software Development Center	163
14.2	ORNL, HTSC Pilot Center Cooperative Agreements	165
17.1	Nikon's "Consortia" R&D	204
17.2	ERATO's Yoshida Nanomechanism Project	205
17.3	ERATO's Kunitake Molecular Architecture Project	206
17.4	Researchers in ERATO by project	207
17.5	Transfer of ERATO research results	208
20.1	Themes and guidelines	234

Preface

Spurred on by increasing international competition, the rising costs of advanced research, the need to leverage scarce scientific and technical talent, and the desire to share the risks associated with technology generation and commercialization, technology companies are banding together in research and development consortia and innovative strategic alliances. Managers in these new types of organizations face the intriguing paradox of competition and cooperation. To compete more effectively in international markets, they must find effective ways to cooperate across research and organizational boundaries.

Consortia and strategic alliances are arrangements among organizations to work together to gain access to technology and markets and to accomplish objectives of mutual benefit. They vary in organizational structure, technological emphasis, funding mechanisms, and personnel make-up. They may be formed to enhance positioning in key markets, gain credibility in a new niche, leverage capital and human resources, expand marketing and distribution capabilities, or build stronger and more direct networks to customers. Yet, they all share one abiding task—managing technology transfer in an efficient, timely manner to accelerate the commercialization of technology and thus speed access to new and expanding markets.

These arrangements pose unique management challenges. Because the members may come from very different corporate cultures, present different managerial priorities, policies, and procedures, and emphasize different and sometimes conflicting objectives, management faces a variety of organizational, technological, strategic, and cultural barriers to transferring technology expeditiously.

This book addresses these barriers, identifies ways to accelerate the technology transfer process, and provides examples of consortia and strategic alliances and their approaches to managing technology transfer.

The book is divided into seven sections. Part I focusing on leadership perspectives on technology transfer, opens with the views of Mark Lieberman of the U. S. Department of Commerce. Three perspectives are provided from the private sector: Grant A. Dove, Chairman and CEO, Microelectronics and Computer Technology Corporation (MCC); Robert Noyce, the late President and CEO of SEMATECH; and Edward A. Miller, President, National Center for Manufacturing Sciences.

Part II focuses on organizational culture and technology transfer. It includes chapters by Bill Curtis, Director, Software Process Program in the Software Engineering Institute, Carnegie Mellon University; Debra M. Amidon Rogers, Senior Management Systems Research Program Manager, Digital Equipment Corporation; Wendell M. Fields, Manager, Software Initiative Technology Transfer Program, Hewlett-Packard; and Michael S. Rubin, President, Molinaro/Rubin Associates.

Policies and procedures for technology transfer is the focus of Part III, with chapters by William M. Evan of the Wharton School, University of Pennsylvania and Paul Olk of the School of Management, University of California-Irvine; Larry Novak, External Technology Manager, Semiconductor Group, Texas Instruments; and Claude DelFosse, Vice President of Technology Transfer, Software Productivity Consortium.

Part IV deals with overcoming barriers to technology transfer. It includes chapters by Raymond W. Smilor, Executive Director and David V. Gibson, Research Fellow, IC^2 Institute, and Graduate School of Business, the University of Texas at Austin; John F. Hesselberth, Vice President, Fibers Research and Development, E. I. duPont de Nemours & Company; and Warren D. Siemens, Deputy to the Vice President of Technology Transfer, Martin Marietta Energy Systems, Inc.

The next two sections of the book provide Japanese and European perspectives on technology transfer. Part V includes contributions from Takashi Kurozumi, Deputy Director, Research Center, the Institute for New Generation Computer Technology (ICOT); Fumihiko Kamijo, Professor at the Information Science Laboratory, Tokai University in Kanagawa, Japan; and Alan Engel, Overseas Representative ERATO, Research Development Corporation of Japan. The contributors to Part VI are Hervé Gallarie, Vice President Software Development, GSI, Paris, France; and Dieter Schütt, Director, Systems and Networks, Corporate R&D of Siemens AG.

Part VII identifies new initiatives for technology transfer. The contributors to this section are Syed Z. Shariq, Assistant Director for Science and Industry, NASA-Ames Research Center; J. Grant Brewen, President and CEO, Biotechnology Research and Development Corporation; and Louis D. Higgs, Senior Fellow, Technology and International Trade, Center for the New West.

Technology Transfer in Consortia and Strategic Alliances links theory with practice. It provides insights by leading scholars and practitioners on ways to facilitate the technology transfer process and increase competitiveness in global markets.

Acknowledgments

Technology Transfer in Consortia and Strategic Alliances is based on the conference "Consortia and Technology Transfer," which was held at the IC^2 Institute at the University of Texas at Austin, in December, 1989. We thank the following sponsors for their support of and participation in this conference: The IC^2 Institute, the Graduate School of Business, and the College of Communication at the University of Texas at Austin; the Microelectronics and Computer Technology Corporation (MCC); the National Center for Manufacturing Sciences; KPMG Peat Marwick High Technology Practice; the Technology Transfer Society of America; the Center for the New West; and the RGK Foundation.

We are especially indebted to the following individuals from MCC who supported and participated in the conference: Grant A. Dove, Chairman and CEO; Jim Babcock, Manager, Shareholder Relations and Technology Transfer, Software Technology Program; Mark Eaton, Director, International and Associates Program; Ted Ralston, International Liaison Representative; and Mary Kragie, Manager of Corporate Development.

We are also indebted to Edward A. Miller, President, the National Center for Manufacturing Sciences; S. Thomas Moser, former Chairman, High Technology Practice, KPMG; and Phillip Burgess, President, and Lou Higgs, Senior Fellow, the Center for the New West.

We also thank Ronya Kozmetsky, President, Cynthia Smith, Director of Administration and Operations, and Milessa Brown and Jami Hampton of the RGK Foundation for helping to organize the conference. We greatly appreciate the support and encouragement of George Kozmetsky, Director, the IC^2 Institute. We also wish to thank Rose R. Orendain of the IC^2 Institute for her assistance with the conference. We are grateful to Linda Teague who prepared the manuscript and provided camera-ready figures and tables for publication.

We thank the following moderators for the conference: George Kozmetsky, Director, IC^2 Institute and J.M. West Chair Professor, the University of Texas at Austin; Frederick Williams, Mary Gibbs Nones Centennial Chair Professor, College of Communication, the University of Texas at Austin; James D. Babcock, Manager, Shareholder Relations and Technology Transfer, Software Technology Program, MCC; James M. Wyckoff, President, Technology Transfer Society; Mark Eaton, Director, International and Associate Programs, MCC; Ted Ralston, International Liaison Representative, MCC; and Kirsten Nyrop, former National Coordinator, Technology Development Services, KPMG Peat Marwick.

We appreciate the support of Dean Robert E. Witt of the Graduate School of Business, and Dean Robert Jeffrey and Professor Frederick Williams of the College of Communication at the University of Texas at Austin.

We thank Rowman & Littlefield for their editorial assistance and direction. We especially appreciate the support and advice of Editor-in-Chief, Jonathan Sisk.

We are grateful to each of the authors of the following chapters for their important contributions to this book. Each of us also wishes to thank his co-editor for making the work on this book such a productive and worthwhile experience.

Part I

Leadership Perspectives on Technology Transfer

1

The Importance of Technology Transfer and Commercialization to U.S. Competitiveness

Mark S. Lieberman

The United States has been experiencing the winds of change in many ways, particularly in the form of tough economic competition from abroad. Considering recent acquisitions of U.S. assets by our Far East competitors, the successful strategies of Pacific Rim countries, the integration of Europe, and the prodemocracy changes in the Soviet bloc, it should be clear that we now live in a global marketplace with never-ending challenges and opportunities. But how well are we doing in that competitive arena? And will we be ahead of the change curve or behind it? The international competitiveness battle will be won or lost through private-sector initiatives. While government at all levels certainly has a role—the private sector will determine the future of the United States.

Continued progress must be made in transfer technology from federal laboratories, universities, and consortia to the private sector. When I refer to technology transfer, I am referring to proprietary technology—which is protected by intellectual property rights—not the technology that is freely diffused to the public domain. While the latter is important, the former gives us the incentive to commercialize. In this chapter I focus on what we must do after technology has been transferred, to assure that the United States remains competitive in the international marketplace. This means efficiently commercializing sophisticated technology in order to produce high-quality products tailored to customer needs worldwide.

It is acknowledged throughout the world that the United States excels at basic research. No other country comes close to the United States in the number of Nobel Prizes for sciences. But there is a growing consensus that we have not done as well in applying that research to the development of new products and processes. The Japanese, by contrast, have focused less on basic research, but have done an exceedingly good job in developing laboratory technology into commercial products and then marketing these products. They have made no apology for using ideas from U.S. labs or elsewhere. In fact, there are numerous examples of foreign products developed from U.S. technology which U.S. companies chose not to commercially exploit themselves.

The chairman of Sony, Akio Morita, summed up his view of our problems in his recent book, *The Japan That Can Say No* (1990). A translated except reads as follows:

> The main reason why Japan's industrial might has become so strong is not that it borrows basic technologies—though, to be sure, many have been brought in from abroad—but that it leads the world in devising ways of creating products derived from those basic technologies. America is by no means lacking in technology. But it does lack the creativity to apply new technologies commercially. This, I believe, is America's biggest problem. On the other hand, it is Japan's strongest point.

Clearly, we are not as strong in commercializing technology as we used to be. However, there is a risk of expecting too much in the way of improvement in U.S. industrial competitiveness from technology transfer alone. Indeed, technology transfer in isolation simply puts the technology where there is only the potential for exploitation. We must take this into consideration in devising our long-term strategies.

During the 1980s, we lost the lead in key industries, including electronics and microchips. As we move into the 1990s, we face the prospect of losing further ground to competitors around the world. Our private sector has become too concerned with short-term profit. Unlike some of our foreign competitors, we exchange wealth through corporate restructuring rather than creating new wealth through innovation. As Mr. Morita stated in his book:

> Americans today make money by "handling" money and shifting it around, instead of creating and producing goods with some actual value.

The Importance of Consortia

The Department of Commerce recognizes that in order to improve our competitive postures and overall economic security, the United States must make evolutionary, yet vital, changes in our economic and technology policy. We must, of course, recognize and take into account all of the factors that affect competitiveness. This includes not only commercialization of emerging technologies, but also lowering the cost of capital to finance new investments in high-technology enterprises, strengthening our research base and work-force skills, improving the quality of our products and manufacturing processes, and focusing on exports to overseas markets.

We must also recognize that the time frame for development of new products has been compressed with the growing use of concurrent engineering and consortia principles by our foreign competitors. Successful commercialization requires simultaneous advances in design, manufacture, finance, environmental soundness, and marketing, as well as cooperation by consortium members responsible for these advances. Our foreign competitors achieve these goals very successfully. In the horizontally and vertically integrated Japanese *kieretsu*, diverse industry groups are poised and ready to exploit research generated by any member of the group. Sharing the development cost and the long term funding for internally financed development projects are advantages these immense organizations have over the competition.

Consortia such as MCC (Microelectronics and Computer Technology Corporation), SEMATECH, and the rest of the over 200 cooperative ventures now registered under provisions of the National Cooperative Research Act of 1984 are at the cutting edge of our response to the competitive challenge. The MCC mission statement captures the essence of the roll of a cooperative venture:

> MCC is a cooperative venture owned by U.S. corporations. [The] mission is to accelerate the creation, delivery, and commercialization of advanced microelectronic technology, by providing . . . participants with useful, timely, and competitive research results [and] helping participants to secure and retain a competitive advantage in markets of their choice.

Our national challenge should be the same. We must broaden cooperation of U.S. firms of all sizes and in all industries in enabling technologies such as advanced materials, digital imaging, microelectronics, flexible manufacturing, genetic engineering, optoelectronics, and high-performance computing. These technologies have the potential to generate a wide array of applications substantially impacting the future of many industrial sectors.

Vertically integrated industrial entities that can cross-subsidize the significant, long-term R&D investment that is required and respond to increasing global demand for customized products with a short life cycle have a strategic advantage in manufacturing and in capturing future markets for the many end products of these technologies. Large-system management approaches are required to develop and commercialize these enabling technologies. Success in developing and bringing these technologies to the market, under the considerable time pressure our competitors create, requires parallel R&D involving many disciplines and firms from different industries, each contributing complementary pieces of the larger system. The nation's success in applying these technologies depends upon the private sector, which must make critical decisions on how much capital, time, and effort to invest. The government, on the other hand, will play a catalytic role.

The Department of Commerce continues the nation's efforts, begun in the early 1980s, to encourage joint research ventures under the provisions of the National Cooperative Research Act. The act provides limited antitrust protection so that firms can pool resources and develop new technologies to compete effectively in the global marketplace. The department is presently looking into various legislative initiatives that would extend antitrust protection to manufacturing, marketing, and distribution cooperatives.

The Department of Commerce is also looking into other ways to promote the formation of consortia. Just recently, Congress appointed $10 million in start-up funds to be used by the Advanced Technology Program authorized by the Omnibus Trade and Competitiveness Act of 1988. These funds are intended to be used to gain operational experience working with industrial consortia which will focus on the development of generic technologies by investment in a variety of modest efforts at different stages in the technology development cycle. Modest investment would offer the opportunity to gain greater leverage on the federal dollars invested by working directly with a variety of large and small companies, and state or local government and even other federal agencies where appropriate.

This program is just one example of the government's increased awareness of technology's importance to the economy and its recognition of an important federal responsibility to assist the private-sector response to the industrial competitiveness challenge. Other examples are the elevation of the president's science adviser to cabinet rank with responsibility broadened to include technology as well as science, and the creation of the new Technology Administration in the Department of Commerce which will act as the lead facilitator among government agencies in seeking to strengthen the capability of the U.S. private sector to commercialize technology developed in U.S. laboratories, whether federal, university, or private. The Technology Administration's mission is as follows:

> The Technology Administration develops and promotes Federal technology policies and programs to increase U.S. commercial and industrial innovation, productivity, and growth through consultation and collaboration with U.S. industrial and nonprofit sectors.

The Technology Administration's three principal organizational units—the National Institute of Standards and Technology, the National Technical Information Service, and the Assistant Secretary for Technology Policy—all work toward this goal.

Other Contributing Factors

Information about U.S. and foreign government-sponsored research is a cornerstone of any long-range competitiveness strategy. The National Technical Information Service (NTIS) is the key agency in this regard, channeling the results of U.S. and foreign government-sponsored R&D to U.S. industry through its publications and seminars. In essence, NTIS is one of the world's leading processors of specialty information. On the other hand, the National Institute of Standards Technology (NIST)—the former National Bureau of Standards—is the only federal research laboratory with the stated mission of directly serving industry. While it specializes in standards and measurements, NIST also performs science and technology research in areas such as semiconductors, optical fibers, computer standards, and biotechnology.

Commercializing technology means more than capitalizing on breakthrough ideas in the laboratory. We must be able to not only develop the better "mousetrap," but manufacture a quality mousetrap more efficiently so that we can compete internationally. Through NIST, the Department of Commerce administers the Malcolm Balridge National Quality Award, which was established to promote quality awareness and successful quality strategies. Announced yearly, this presidential-level award has been very successful in putting quality on the national agenda by focusing companies on "quality" as an important factor in international competition.

An integral part in any quality initiative is human resources—perhaps our most precious resource and the key to our future competitiveness. Only by properly educating our children, training our work force, and retraining management can we assure the efficient commercialization of advanced and future technology. Although education is primarily a state and local government responsibility, every blue-ribbon panel that has examined the subject in recent years has called for the business community to get involved. This includes getting involved at all levels in the local schools, in curriculum development, and in the community as role models and volunteers. Firms help education fulfill a need, not just for funds, but with their special technical expertise and knowledge of what we need and what works. High-tech firms such as IBM are well known for their interest in education. The recent $30 million gift by RJR Nabisco to North Carolina schools is another example. This gift is not simply a contribution, it is aimed at changing the way the schools operate in order to produce the work force we need to remain competitive.

The Department of Commerce

Within the Technology Administration, the Office of the Assistant Secretary for Technology Policy develops, coordinates, and advocates policies, including legal initiatives to promote U.S. competitiveness through technology commercialization. This also includes coordinating international policies and programs to ensure that intellectual property rights are equitably distributed in international agreements and that they are consistent with technology transfer policies for universities and federal labs.

The Department of Commerce also interfaces with the private sector and other governmental agencies on technology transfer from U.S. laboratories to U.S. industry. In this regard, the department is responsible for oversight and implementation of the Federal Technology Transfer Act. The act was passed in 1986 to improve the flow of technology from federal labs to the private sector by decentralizing authority to labs in order to allow them to work more closely with private firms. Previously, private firms generally had to negotiate with agency headquarters in Washington. This was a lengthy and inefficient process because the private firms did not deal directly with those most knowledgeable about the specific scientific development. In general, federal agencies were very poor at commercializing their laboratory technology. The Packard Commission in 1983 estimated that agencies owned 28,000 patents but licensed only three to five percent of them. In addition, government royalty income was only about $2 million per year on annual expenditures of $20 billion to run the laboratories.

The new Federal Technology Transfer Act permits the private sector to enter into cooperative R&D agreements after negotiating directly with the laboratories where the research is being conducted. Both parties can bring people, equipment, and facilities into the cooperative R&D agreement, while the private firms can contribute money. The laboratory can even grant future rights to the technology developed under the agreement. The act requires federal lab directors to give special consideration to small businesses and business units located in the United States that agree to manufacture in the United States. The act also provides commercialization incentives in the form of royalty payments to federal scientists and the laboratory itself.

The Federal Technology Transfer Act has already had an impact, and the labs have entered into over 200 cooperative R&D agreements. Most agencies report a substantial increase in invention disclosures and patent applications. An example of how the private sector can utilize these statutes is the recently announced superconductivity cooperative R&D agreement between the Department of Energy's Los Alamos facility, DuPont, and Hewlett-Packard. The joint technical program, which will employ 25 or more researchers at a cost of $11 million over a three-year period, will focus on the development of thin-film high-temperature materials for electronic components.

The Department of Commerce also oversees commercialization of technology developed by universities from federally funded research. Our role in this regard is to assist other federal agencies in complying with the provisions of the 1980 Bayh-Dole Act. This statute gives nonprofit entities, primarily universities and small businesses, the right to retain patents for technology developed with federal funding. Universities, which are the major beneficiaries of this legislation, can license companies that agree to manufacture products resulting from federally funded inventions in the United States. Passage of the Bayh-Dole Act has stimulated university research and strengthened university relations with business. Since 1980, private-sector funding of university research has more than doubled. In addition, the number of patents issued annually to universities has increased threefold, and royalties from patent licenses have also risen sharply for many universities. For example, Stanford University's royalty income grew from $655,000 in 1980 to $11.5 million in 1988. Also, a growing number of start-up firms utilizing university research are being established—MIT alone spun off eight start-ups in 1988.

Despite the improvement in federal technology transfer, as a nation we are not doing as well as we could or should. Almost all concerned participants and observers feel that we have only just scratched the surface and that federal laboratories could be making much greater contributions to the overall national well-being and competitiveness through technology transfer. While we continue to address the impediments to federal technology transfer that the secretary of commerce identified in his first biennial report to the president and Congress, we have found that perhaps the largest impediment to commercialization is the failure of the private sector to utilize the Bayh-Dole and Federal Technology Transfer Acts.

Conclusion

Federal statutes have unlocked the door—now it is up to the private sector to open it. But in a few cases, just the opposite has happened, and some university technology transfer managers have even confided that they welcome foreign competitors because of their financial support of the university's research programs and because foreigners are more knowledgeable and foreign business easier to negotiate with than their U.S. counterparts.

Clearly, the civilian sector must take the lead in commercializing university and federal lab technology. While the government plays an important role, it is the private sector that will determine the future of U.S. competitiveness. Individual firms and trade associations would be well advised to assess their weaknesses and shop the federal technology marketplace for technology they might use. In short the private sector must "pull" technology out of the laboratories instead of relying on the federal labs to market their own R&D.

The Department of Commerce is doing its part to explore what it can do to assure that the private sector has maximum opportunity to commercialize technologies developed with federal funds. We want to find means by which the private sector can interact more closely with universities and government laboratories to utilize the tremendous amount of technology that is being developed. The task will not be easy, because in many cases the private sector, academic, and federal research cultures are so different. Seeking to understand and overcome these impediments, the Department of Commerce has worked very closely with its sister agencies. For example, we learned that some firms would not enter into cooperative agreements with agency labs because they were concerned that their proprietary information could be obtained by the competition pursuant to a Freedom of Information Act (FOIA) request. There now exists legislation, signed by President Bush, that protects proprietary information from disclosure by providing a five-year FOIA exemption for jointly developed information.

While such legislative changes are important, the United States must recognize that if the nation does not have the manufacturing capability, then technology transfer is incidental. With this in mind and recognizing that most manufacturing in the United States is performed in over 340,000 small businesses, the Department of Commerce has encouraged small firms to adopt the shared flexible computer-integrated manufacturing (FCIM) concept. This manufacturing technique integrates off-the-shelf technology, such as robotic machine and handling tools, with computer-aided design (CAD) and computer-aided manufacturing (CAM) so that a variety of parts, in small batches, can be efficiently manufactured. The shared center leases time on the manufacturing system to small businesses so they can adopt automation capabilities without the risk and enormous capital investment required.

We have found two major obstacles that firms face if they wish to use the shared center. First, traditional business practices, such as management, financing, and accounting principles, do not readily apply. Second, many of these firms are not familiar with how automation technology works and how it can be employed. To overcome these obstacles, the Department of Commerce assists businesses and universities in addressing their concerns. We also strive to diffuse advanced manufacturing technology to industry through our Advanced Manufacturing Research Facility and the Regional Manufacturing Technology Transfer Centers in Troy, New York; Columbia, South Carolina; and Cleveland, Ohio. We are also intimately involved with the several dozen shared FCIM centers which are in the planning stages and will be financed by the private sector and state governments. These centers act as teaching facilities.

State and local governments have recognized technology's potential to create new industries and new jobs and to strengthen local economies. Many are establishing innovative ways of linking government, university, and industrial technology programs. The Department of Commerce seeks to encourage and assist these efforts through the establishment of the state and local clearinghouse programs. Statutorily created in 1988, the clearinghouse will administer a database on federal, state, and local initiatives to commercialize technology, as well as have personnel who can help the state and local governments exchange information about their technology promotion efforts and analyze what works and under what conditions. The Congress and the Department of Commerce's sister agencies within the federal government will also have access to the clearinghouse.

Having reviewed what the Department of Commerce is doing to support the nation's quest for global competitiveness, I want to be sure our perspective is clear. The important aspect of all of our initiatives is that they should be led, not by the federal government, but by private industry with growing involvements and support of state and local governments. We in the federal government can only try to set the climate for private commercialization of technology. In the end, it is the private sector's will and capabilities that will determine the future standing of the United States in the global marketplace. With smarter private and public sectors, and better public-private cooperation, we can remain ahead of the changes now occurring and be more in control of our own destiny.

Reference

Morita, Akio, *The Japan That Can Say No.* 1990.

2

Encouraging Innovation and Excellence in Technology Transfer

Grant A. Dove

Over the past several years the United States has gotten a rude awakening regarding technology transfer. The country has found that although we enjoy a tremendous technology resource base, we have not been utilizing these resources effectively. I believe consortia are national resources that are changing the United States technology and business culture. In their report on technology transfer at MCC, Smilor, Gibson, and Avery (1990) state,

> Technology transfer has proven to be a complex, difficult process for many U.S. firms, even across different functions within a single product division. Getting innovative ideas from the lab through production, marketing, and sales to the customer in a timely, profitable manner is proving to be a significant challenge, even for the best managed U.S. firms.

Getting basic science from our U.S. labs to the customer in a timely and profitable manner involves a lot of people with very different skill sets. You have got to have excellent researchers; you have got to have people that understand industry problems; you have got to have engineers who will develop the technology from the labs; and you have got to have marketing folks who can take that technology to the customer.

It is interesting that management runs into the same types of problems in a consortium as it does in a company—challenges such as the "Not Invented Here" (NIH) syndrome and technology that is too big a jump from today's mainstream. Engineers and people involved in product-line development are often afraid of revolutionary advances. They want technology they can learn right away and apply to solve their current problems, or enhance their current product lines—a 15 percent to 25 percent change rather than a 75 percent to 100 percent change.

The Microelectronics and Computer Technology Corporation (MCC) can be viewed as a "technology conduit." We create new technology and pass it to our participants, but we also see a tremendous technology pool that already exists in the United States. MCC has created new alliances with universities, government agencies, and labs to make sure that if the technology these organizations develop can be of benefit to our participants we incorporate it into our projects. We also think we can help universities and government agencies make decisions on the priority and relevance of research due to the unique perspective our 37 member companies provide on problems and needs. Thus, we see MCC as a potential two-way "conduit" for universities and national labs.

At MCC, we have two key challenges:

- identifying technology gaps and conducting and importing excellent R&D;
- transferring the technology to our participants and helping ensure that the technology adds value to the company.

The following four topics relate to these challenges and to encouraging innovation and excellence in technology transfer. After a brief description of MCC, this chapter will focus on managing research in a vertically integrated consortium. The chapter will then present MCC's view of technology transfer and offer some guidelines, approaches, and lessons learned.

MCC Overview

MCC is a cooperative enterprise whose mission is to strengthen and sustain U.S. competitiveness in information technologies. Our objective is excellence in meeting broad industry needs through application-driven research, development, and timely deployment of innovative technology.

MCC's participants (shareholders and associates) represent a diverse group of outstanding materials, semiconductor, computer, aerospace, manufacturing, and telecommunications companies. MCC's 21 shareholders include Advanced Micro Devices, Inc.; Bellcore; Boeing Company; Cadence Design Systems, Inc.; Control Data Corporation; Digital Equipment Corporation; Eastman Kodak Company; General Electric Company; Harris Corporation; Hewlett-Packard Company; Honeywell, Inc.; Hughes Aircraft Company; Lockheed Corporation; Martin Marietta; 3M; Motorola, Inc.; National Semiconductor; Northern Telecom Limited; NCR Corporation; Rockwell International; and Westinghouse Electric.

MCC also has 24 associate members, including Apple, Sun Microsystems, E-Systems, Tracor, Dell, and TRW. Some of these associate companies fund research at MCC, and some monitor the research at a nonproprietary level.

MCC's participants are frequently competitors in the marketplace. Yet they work closely together at MCC and share a common goal—to create the technology necessary to make computer applications and processes faster, more reliable, and capable of performing more complex tasks at a higher level of quality and at a much lower cost. MCC conducts research in five main areas: advanced computing technology; computer-aided design; high-temperature superconductivity; semiconductor packaging/interconnect; and software technology.

MCC's intellectual environment is not limited to just our shareholders and associates. MCC views itself as a conduit between industry and our national intellectual resources, such as universities and national laboratories. For example, MCC works closely with the University of Texas in many research pursuits, such as packaging and software technology, and with Stanford University in the area of artificial intelligence. We have also created a unique consortium with the Texas Center for Superconductivity at the University of Houston, which matches our superconductor industrial process and device fabrication expertise with Houston's outstanding materials research. This consortium also includes the participation of the three National Lab Superconductivity Pilot Centers.

Vertical Integration

A key challenge facing the information industry is identifying, developing, and commercializing technology that is critical in guaranteeing a competitive stance in world markets. MCC works hard to understand the diverse cultures, management styles, and organizational structures of its participants. Because of the unique technical problems MCC is addressing, we have attracted materials and systems companies and component manufacturers. Management and researchers at MCC must understand the participating companies and the role of the consortium within each company to ensure that we are providing the research and the technology transfer service that each company requires. MCC is continually evaluating its effectiveness in:

- Understanding the research process and the research infrastructure, which includes industry, government, and university resources;
- Articulating a vision: what is the problem or the technology gap the industry faces and how will MCC help solve it;
- Scheduling deliverables and breakthroughs. MCC's research plans focus on the development of evolutions on the way to revolutions;
- Maintaining the ability and the flexibility to respond to new opportunities, such as exploiting an invention;
- Managing its staff well—balancing the mix of technical brilliance and client service.

By virtue of its diverse membership, MCC has created a vertically integrated environment. MCC views the essence of vertical integration as the orchestration of R&D and investment. This is a powerful approach to solving industrial problems and implementing solutions, but it also can create challenges. The companies often do not have a common view of the technology, and technology transfer is more complex as a result. A real strength of vertical integration is the correlation of the "views of the elephant by the blind men" into a common agreement on the problems needing solutions and the priority of attacking them. This can include technology, process development, implementation equipment, or a prototype manufacturing capability. However, in the longer term this vertically integrated structure may be very important as we continue to address issues of U.S. competitiveness because:

- Most U.S. companies are not vertically integrated;
- Vertically integrated foreign companies are getting stronger;

- Global competitiveness demands cooperation between industry segments.

In some MCC research programs our membership emulates large foreign competitors because of the blend of suppliers, vendors, and users. We have certainly seen the strength of this approach in our CAD program, where companies want software support. Since this is not really the mission of MCC, we now have CAD vendors as members supporting the technology MCC is developing.

Approaches to Technology Transfer in a Consortium

To be creative in technology transfer, we must understand our participants and we must also recognize the principles that govern the process of technology transfer:

- All of MCC's companies—our customers—are different;
- All companies are equally important and must feel we are treating them equitably;
- All must be satisfied, and achieving this result might require trying several different technology transfer approaches;
- All must recognize the importance of their role as receiver of the technology.

I believe there is really an art and a science to technology transfer. The science of the process is a bit easier to get your arms around. It is the technical documentation, the drawings, the specifications, the process description, the source code. It is the training involved in moving technology into a new environment. It is the support and follow-up in the form of hotlines, updates, and fixes.

Keeping up with these challenges is difficult enough, but they are only part of the story. The art of research commercialization is where the intellectual entrepreneur must use his or her skill to ease a technology from the lab to the user to production. If an innovation is going to make it, it must have behind it an individual who believes strongly in its value, understands its potential applications, and has the energy and conviction to communicate his beliefs. Such an individual must play the role of an outspoken champion—creating excitement and motivating the people who will receive and ultimately use the technology. The champion must care intensely about the technology, "preach the gospel" on why it is important, and must not take no for an answer. Such champions are the people that make the system work.

MCC recognizes the importance of champions and rewards their success. MCC has an attractive incentive program, which has as one of its components an evaluation of an employee's effectiveness in transferring technology. Our bylaws also outline MCC's philosophy that the individuals who participate in the creation and transfer of licensed technology will receive a portion of the royalty stream that technology creates for MCC. We are now beginning to see some individuals reaping the monetary benefits of their work at MCC.

Most importantly, technology transfer incentives cannot end at MCC's doors. The companies that participate in a consortium must also put in place similar incentive systems to encourage the technology users to quickly and effectively utilize the technology in which their company is investing.

Throughout the technology transfer process the artistic and the scientific principles are at work. During MCC's seven years of operation, we have seen some approaches to technology transfer work very well, while others have been less successful. MCC researchers are continually encouraged to try new approaches to move the technology into the participating companies. Once a technology has been identified and is being developed at MCC, there are many different ways that it can move into the participating companies.

- Ideas and research move into company shadow programs.
- Technologies are transferred through formal transfer packages and training, and then supported or manufactured by MCC members or licensees.
- Co-collaborators in participating companies work with MCC researchers, but in their company labs.

Ideas and Research to Participating Company Shadow Projects

In many cases, as a technology is being developed at MCC, participating companies begin complementary research and development programs based on MCC's work. This allows the company to closely monitor MCC's research and validate research results. It requires early commitment and investment by a company, but this approach can reduce risk, improve quality, and accelerate progress. This is often the strategy selected by a market leader. This was the approach taken by Digital Equipment Corporation in their use of MCC's Tape Automated Bonding (TAB) semiconductor manufacturing technology. DEC began integrating TAB technology into its design and production process in 1985 and 1986 and provided valuable feedback to MCC researchers. DEC's use of TAB in its VAX 6000-Model 400 computer in 1989 makes them among the first American companies to use TAB in volume production.

Technology Transfer Packages and Support

As a technology matures at MCC, technology training workshops and seminars are held. These meetings give scientists and engineers from participating companies the opportunity to meet with MCC researchers to learn how to use and apply MCC technology. In some cases, technology transferred to companies is readily integrated into existing or planned hardware or software systems. In other cases, recognizing that MCC cannot do all that they would like, participating companies have suggested other ways of transferring and supporting the technology. In other words, they have looked for ways to meet a technology transfer need, but still "keep it in the family." For example, in the case of MCC's Computer-Aided Design program, participating companies were enthusiastic about the CAD software tools we were developing, but were concerned about long-term software support of the tools once they were transferred from MCC. This issue has been addressed recently by the participation of CAD vendors in MCC. These companies, such as Cadence, will provide long-term software maintenance and allow MCC to continue developing new CAD tools.

In MCC's Packaging/Interconnect program, a similar set of issues arose early in 1989. MCC had developed an advanced technology for bonding chips to substrates, called laser bonding. MCC participants chose not to manufacture the bonder, but wanted the mature technology transferred to their companies. MCC licensed the technology to Electro Scientific Industries (ESI) who will manufacture the system initially for MCC and its participants and then make it available to the general industry.

In both of these cases, MCC and its participants created new approaches to ensure that MCC technology moved as quickly and efficiently as possible into the participating companies. However, these precedents have further reaching implications for technology transfer: non-MCC shareholders will also have access to advanced U.S. technology.

Co-Collaborators

One of the newest approaches MCC is taking in the technology transfer process is in one of its potentially revolutionary projects. Researchers in the Optics in Computing project recognize that revolutions are sometimes hard to move into companies. Therefore, the MCC optics project is working hard to create "buy-in" partnerships to the technology. The purpose of establishing co-collaborator relationships with each participating company is to encourage scientists and engineers to think about how they will use the technology in the long run. MCC believes the co-collaborator approach increases the probability that MCC technology will be deployed by participating companies because:

- Shareholder researchers are directly involved, thereby increasing their ownership of technology;
- Novel research applications are increased by increasing the number of independent projects;
- Acceptance and visibility of the technology will increase due to on-site demonstration projects within companies.

There are no rules and regulations for these or any new technology transfer approaches. MCC recognizes that at times a particular technology can drive a transfer approach. Maintaining flexibility is essential in allowing creative approaches and encouraging researchers to try out new ideas. MCC has actually gone through several phases in its technology transfer evolution. An initial phase might have been termed the "Here it is. Come get it" phase. As MCC and its relationships with its participants matured, its technology transfer efforts became much more active and interactive, and formal transfer processes like meetings and training workshops became much more frequent. We have now entered a third phase in our development: collaboration. This personalized approach can be more complex, but it also can be very effective. During the past six years, MCC has learned that a consortium must get the attention of its participants. This means the consortium must actively market its technology to its owners. MCC has known that a consortium must have credibility and integrity; however, building these relationships and gaining a company's respect takes time. But a consortium must take the time to achieve these goals if technology transfer is to be successful.

MCC acknowledges that having an authoritative voice within participating companies can truly aid the technology transfer process. For example, one of our shareholders said that MCC has had a "unifying" effect on the R&D and product divisions within the company in that the consortium helped drive an accepted approach across the company. It is interesting to see not only technology but also culture being transferred from MCC. However, such results take time. A consortium must have time to make the evolutionary advances and successes in order to move toward the revolutionary advances. A technology at MCC is only successful if the participants can use it. MCC has also learned that successful technology transfer includes bolstering its role as a technology conduit. We represent a very unique resource to our participants and have the ability to "fill a gap" between them and university and national labs, by accessing and "digesting" technologies they might not normally acquire.

Reference

Smilor, R., D. Gibson, and C. Avery. "R&D Consortia and Technology Transfer: Initial Lessons from MCC," *Journal of Technology Transfer,* vol. 14, no. 2, Spring 1990.

3

SEMATECH: Necessary but Not Sufficient

Robert Noyce

World semiconductor leadership is generally measured by the ability to make integrated circuits with smaller and smaller dimensions. Today's most advanced products use circuits about one micron wide—that is about one-hundredth as wide as a human hair. Future dimensions will be much smaller. SEMATECH hopes to be producing devices with circuits no wider than one-half micron by 1992 and .35 micron by 1993. Volume production at these levels will reestablish U.S. leadership and produce products with four times the capacity of today's devices. In SEMATECH's three-phase operating plan, Phase 1 aimed at establishing a one-micron baseline, which we have done to our satisfaction. Phase 2 will develop half-micron capability, and Phase 3 .35-micron technology.

While SEMATECH works toward these goals, however, we have grown very concerned that many U.S.–based suppliers of equipment and materials are vanishing—through mergers, foreign acquisitions, or just going out of business. This development threatens SEMATECH's mission and the U.S. semiconductor industry. Therefore, our immediate challenge is to strengthen this vital infrastructure of suppliers even as we work on the technical aspects of our mission.

To achieve both, we have aggressively initiated joint development contracts or equipment improvement programs with a number of suppliers. Our 1989 external development spending will be about $108 million and will exceed $120 million in 1990. Some of the projects include:

- an advanced, laser-based system to make master images for printing integrated circuits;
- the next generation tool for printing circuit patterns into silicon wafers;
- a state-of-the-art system for quickly finding wafer defects.

SEMATECH recently achieved a significant milestone by making and measuring the first Phase 2 devices, with U.S.–made equipment. We did this just

24 days after getting tools from our suppliers that made these accurate measurements possible. It is worth noting that we have begun work on Phase 2 items well ahead of schedule.[1]

Of course, none of this hard work has any meaning without technology transfer. It is important to note that SEMATECH is not making chips for sale—we acquire knowledge that we transfer to our member companies and the Defense Department who then put that knowledge to use. One of the most significant steps we have made at SEMATECH, however, is not something you can hold in your hand or write down on paper. We feel we have begun the important process of helping to change attitudes.

We are a precompetitive consortium that demands from its members a willingness to share knowledge and information with their competitors. But our members—and their suppliers—have come to understand that sharing precompetitive information enables them to be even fiercer competitors, not only among themselves but with the best companies worldwide. Member companies can tell their suppliers more about what they need, what suppliers must do to meet those needs, and how their products fit into the overall process. This kind of information flow does not weaken competitive positions, it makes them stronger. Being an agent to help make this exchange happen has been very gratifying for all of us at SEMATECH. We are beginning to see a significant change in the relationship between suppliers and manufacturers—a change from short-term adversarial attitudes to long-term cooperative efforts that will benefit all concerned.

The Importance of Technology Transfer

Following are some brief comments on the importance of technology transfer at SEMATECH and on the consequences of failure in this area. It is common knowledge that the United States has lost its lead in the worldwide sale of semiconductors. In 1980, the United States produced the majority of semiconductors sold around the world. The Japanese passed the United States in 1986, and the gap continues to widen. Some theorists would argue that it does not matter whether the products you buy are made in Taiwan, the United States, or Libya. Their argument is that control of the intellectual process—the information that went into designing and creating the product—is what matters. Where the finished product is made, and who makes it, is insignificant. I must tell you, I could not disagree more strongly. Manufacturing capability, especially semiconductor manufacturing, remains central to this country's economic and strategic well-being.

The armed forces of the United States grow more dependent on electronics with each passing day. We as consumers are developing the same dependence. Almost every sophisticated consumer product we buy in the 1990s contains some sort of semiconductor-driven device—television sets, VCRs, even our clocks, watches, coffee makers, and telephones. Most of these products are imported, and most of the chips that go into them were made somewhere else, too. If this trend continues, it has serious implications for our national standard of living. There was a time, not so long ago, when our supply of a vital material—crude oil—was curtailed by foreign suppliers. The resulting disruption of our daily lives was huge and frightening. Well, semiconductors are the crude oil of the information age, and the more dependent we become on foreign sources, the higher our risk of an interrupted supply.

To take this point one step further, recent events—especially in Eastern Europe—are making it obvious that the definition of national security is changing. Today, national security is increasingly a factor of economic strength, rather than pure military strength. The sharp distinctions contained in the classic "guns vs. butter" argument are rapidly becoming blurred.

If SEMATECH does fulfill its mission and the United States regains world leadership in making semiconductors, it will only be due to the effective transfer of knowledge to our member companies and the Department of Defense. This is the critical step in the partnership endeavor that SEMATECH represents.

Conclusion

Even though I believe deeply in the importance of the SEMATECH mission, SEMATECH alone is not going to solve the problem. SEMATECH is necessary, but not sufficient.

When we succeed at SEMATECH, as I am confident we will, it would be a terrible shame if this country did not have a work force of flexible, well-educated people to take advantage of the new jobs that will be created. It would also be a shame if we were not able to sell our semiconductors overseas, to help balance our trade. And it would be a shame to reestablish world leadership in this critical area only to find that investment capital in this country was so dear that growth had become impossible.

The United States today does not provide the proper economic environment to let our companies compete. The playing field is not level. Let me hasten to say that I and all of us at SEMATECH, believe in free enterprise. It is the system that made us the richest, most-envied nation in the world. But while our companies have the responsibility to compete, our government has an equally important responsibility to create and maintain an environment in which free enterprise can work—the right soil in which our economic system can grow and come to flower.

In this regard (maintaining the proper environment), our government is not competitive with other governments, such as that of Japan for example. The savings rate in this country is pitiful. It has averaged less than one-third the Japanese rate of savings for the past eight years. The result: Japanese investment capital is plentiful and patient, while American capital is scarce, expensive, and nervous. Corporate executives are forced to manage for the near term because that is as far as their bankers, creditors, and shareholders will let them plan.

The educational system in this country needs repair. Recent scores from standardized 12th-grade tests given in a dozen countries showed U.S. students placed no better than eighth in any of the ten categories tested. Japan finished no worse than third.

Our trading policy is ineffective. We have agreed to buy the products of almost every civilized country in the world without demanding that they be equally willing to buy from us. Again, using Japan as an example, we have been buying more goods from Japan than they have bought from us at the rate of $50 billion a year for the past five years. And how have we been paying for those purchases? Not by selling manufactured goods, because Japan will not buy those from us—they prefer to buy at home. So, we have been selling them real estate, buildings, companies, and technology. Our accumulated national wealth is rapidly being drained overseas. To put it bluntly, we are in an economic battle for survival. In an interesting book called *The Chip War*, the author states:

> What is at stake in the chip war is more than just the loss of another strategic industry . . . the very future of America as a great nation may be the ultimate prize.[2]

I also believe we are in an economic war, and the economic well-being of our nation is squarely in harm's way. We know what we must do: the question is whether we have the will to do it.

Notes

1. On August 25, 1990, SEMATECH announced that the consortium was producing chips with .5-micron widths with U.S.–made equipment.

2. *The Chip War,* Fred Warshofsky, New York: MacMillan Publishing Company, 1989.

4

It's Not Over until It's Transferred!

Edward A. Miller

To reassert ourselves as a globally competitive nation, the United States needs to have world class manufacturing and service organizations and they are going to have to be profitable. There is just no other way.

You would think that with all we have invested in the last 40 years every organization in the United States would be world class and profitable. Just look at the statistics. Not counting 1989 and 1990, we have won 147 Nobel and Fields prizes, 70 percent of them since World War II. The second place country, Great Britain, has 65 to its credit. And our strongest trading partner, Japan, has only 5. So the answer to our competitiveness problems is not exclusively doing great research. What I glean from these and other competitiveness statistics is that as a nation we are not making the right blend of investments and as individual companies we are not addressing the right activities. We are concerned with the flashy items, "the scientific breakthroughs," and have neglected the glamorless items, "the commercial applications." This must change. We must redirect our misplaced attention from creating Nobel laureates to creating noble product and process engineers.

As industry, government, and academic leaders, we are responsible for the decisions and actions of our past. We cannot go back in history and make changes in what we have done, but we can go into the future and make history with what we change. Technology, management, and culture represent three areas where we can transfer better ideas than those we currently use.

Technology

The easiest and most obvious place to begin is with technology. In the last 40 years the United States has spent more that $3.5 trillion on its creation. That is a lot of money, and you would expect that we could deal effectively with change, that we would have standardized essential interfaces, and that we would have all the product and process technology necessary to be competitive. However, this is not the case. For years in our companies we have lived with the axiom, "If it isn't broken, don't fix it." For all too long this has caused us to look some other way when new and promising innovations were presented to us. As a result today most of our companies do not employ anywhere near the best of current practice, much less advanced state-of-the-art technologies and practices. We have failed in the transfer process. We must graduate from a technology environment where change is unusual to an environment where technology change is constant and necessary.

For years our companies have been taught, "If you want to compete successfully, you must differentiate your products." For all to long this has caused us to forego standards and to fight their implementation. As a result, today we bear tremendous cost burdens to develop and maintain proprietary systems and interfaces, while our trading partners have standardized and reduced costs. We have failed in the transfer process. We must move from a standardization process that is slow and adversarial to one with an equal emphasis placed on product and process.

For years we have believed, "Design a better mouse trap, and the world will buy it." For all too long this has caused us to focus on product technology. As a result, today, many of our companies cannot competitively produce the products they design. We have failed in the transfer process. We must migrate from a system that places the emphasis on product exclusively to one with an equal emphasis placed on product and process.

Management

Second, we must consider the area of management. In the last 40 years we have gone from companies with management run by the seat of their pants to companies with the most highly educated management in the world. They are better educated when they start, and each year we spend more dollars on upgrading, training, and reeducating our work forces, including management, than are spent by the federal, state, and local governments combined. That is a lot of money and one would expect that we could create more great companies, develop larger markets, and produce any product competitively. However, this is not the case.

For years our companies have operated with one thought in mind, "Wall Street is expecting a big profit this quarter, so we better give it to them." For all too long this has caused us to dwell on short-term time horizons. As a result, today many of our best executives concentrate on breaking up and destroying their companies, not building them. We have failed in the transfer process. We must do away with the fascination for increasing quarter-to-quarter market profits and begin profiting from quarter-to-quarter market-share increases.

For years our companies have been told, "You can't make a profit in consumer commodity products." For all too long this caused us to cast aside numerous technologies with vast market potentials. As a result today, our trading partners dominate entire industries, and questions are asked daily about our ability to reenter these markets. We have failed in the transfer process. We must revise our view from consumer commodities that generate low profitability to commodities that create great volumes of consumers.

For years our companies have held, "It is easier and better to retreat from highly competitive markets." For all too long this influenced us to constantly redirect our companies, withdraw from markets, and leave unresolved the real competitiveness issues. As a result, today our trading partners have taken over markets where we once were strong, and they are continuing to pursue additional areas of interest. We have failed in the transfer process. We must alter our course from one that abandons products and weakens corporations to one that strengthens corporations and defends markets with better products.

Culture

Perhaps culture is the area of greatest importance for the future. Without question, we have developed the greatest nation of entrepreneurs in the world's history. Their resourcefulness has led to an unprecedented rise in the standard of living enjoyed around the globe. We would expect the best of relations within our companies, a spirit of alliance across industries, and a superb camaraderie throughout the nation. However, this is not the case.

For years our companies have accepted the tenet, "The best way to do everything is my way." For all too long this has driven a wedge between the various layers within our organizations and across all disciplines and departments. As a result, today we expend significant resources devising rules or procedures and finding ways to get around and compete with each other. We have failed in the transfer process. We must move from organizational cultures where we compete to fight each other to cultures where we fight together to compete.

Conclusion

For years in our industries the cliché has been, "We don't need anyone else, we can do it alone." For all too long we have treated our customers, our suppliers, and our competitors as adversaries. As a result, today we have weakened our industries to the virtual breaking point, we have lost precious capabilities, and we have slipped far behind many of our trading partners. We have failed in the transfer process. We must change our industries from cultures where our peers are our most-feared enemies, whom we must fight, to cultures where when we must fight, our peers are our best allies.

For years in our nation we have always known, "government, industry, and academia don't mix." For all too long we have followed a path of distrust in, distaste for, and distance from each other. As a result, today we have brought our nation to its knees with excessive regulation, poor-quality products, and an illiterate work force while our trading partners have grown stronger. We have failed in the transfer process. We must continue changing our nation from one that is built on distrust and maximizes working against each other to one that maximizes trust and is built on working together.

Part II

Organizational Culture and Technology Transfer

5

Technology Transfer in Knowledge-Intensive Organizations

Bill Curtis

Traditionally, technology transfer is thought of as convincing management to install new technology, with the primary issue being the cost of equipment. Although this scenario may still be true in industries such as farming and manufacturing, this chapter will describe some surprising observations about the costs of technology transfer in knowledge-intensive businesses such as software development. Managers in the software industry still think of technology transfer costs as being primarily new capital equipment and expensive software. However, the real technology transfer cost in knowledge-intensive businesses is labor, with the most prominent cost being the time required for human learning. To illustrate this theme, I will describe software-development data that characterize the challenge MCC's Software Technology Program faces in transferring technology to our participating companies.

There are seven attributes that characterize work in knowledge-intensive organizations. These attributes will be described with examples from the domain of software development.

- Large individual differences in performance
- Integration of multiple knowledge domains
- Continuous technological change
- High percentage of work spent learning
- Importance of rationale
- Heavy communication overhead
- Hard-to-measure productivity, quality, and progress.

Software Development as a Knowledge-Intensive Business

Large individual differences in performance. The individual differences in software-development productivity go far beyond those observed in any other field of human endeavor, especially athletics. The difference between the average adult running 100 meters and Ben Johnson's steroid-induced 100-meter world record at the 1988 Olympics is probably not more than two, or at worst three, to one. We regularly see twenty-to-one differences among software-development professionals.

For instance, Figure 5.1 displays data that appeared on the front cover of Barry Boehm's landmark book, *Software Engineering Economics* (1981). The horizontal bars represent the productivity factors that turn into cost drivers in their cost estimating equations for TRW's Defense and Space Group. Notice that the single biggest cost driver by a factor of two is the capability of the personnel and team assigned to the job. It has twice the impact of the complexity of the program they are assigned to build. It is extremely disappointing that the impact of modern programming practices and software tools is miniscule compared to the enormous impact of the team's capability. The impact of technology has often been disappointing in a knowledge-intensive business like software development.

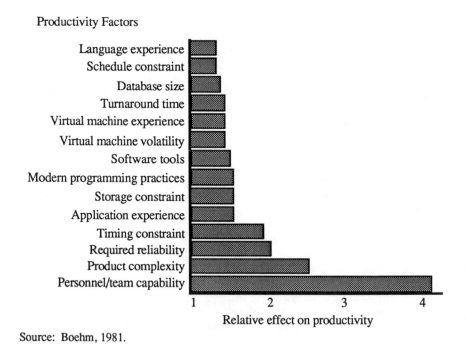

Source: Boehm, 1981.

Figure 5.1. Productivity factors in TRW.

The data in Figure 5.2 describe individual productivity results (computer instructions per hour) on Fortran programs of from ten to several hundred thousand lines of code developed in NASA's Software Engineering Laboratory. For projects of less than 20,000 computer instructions, we observe a 20:1 difference in individual productivity. The contractor NASA uses on these projects puts their inexperienced people on smaller projects, and it is among these less experienced programmers that the greatest range of differences occur. For larger projects (>20,000 source lines of code) the range is reduced to 8:1, but this is still almost an order of magnitude in performance among experienced professionals. We observe these kinds of differences consistently across the software-development industry and suspect that they are typical of other knowledge-intensive businesses as well.

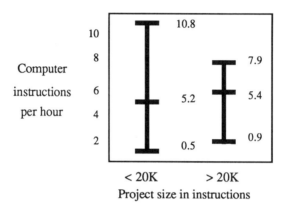

Source: Valett and McGarry, 1989.

Figure 5.2. Individual differences in NASA software.

Integration of multiple knowledge domains. In a study of the design process for large software systems, Curtis, Kranser, and Iscoe (1988) identified the thin spread of application-domain knowledge as a major problem affecting the productivity and quality of systems development. On one out of every three projects they studied, they ran into a guru, usually a senior systems engineer who had dramatic impact on the design and execution of a project. What set these exceptional designers apart was their extremely deep understanding of the application domain (e.g., avionics, telecommunications, electronic funds transfer, etc.). Their extraordinary contribution came in their ability to translate between behaviors in the application domain and structures in domain of computation (i.e., computer science knowledge). Thus, they were able to integrate knowledge from multiple domains much more effectively than their colleagues whose knowledge of the application was thin.

Creation of intellectual artifacts requires professionals to integrate information that comes from multiple knowledge domains. For instance, in building avionics software for jet fighters, there are numerous knowledge domains you would have to integrate before writing successful programs. In the technology domain, you would need to know the Ada language and an assembly language, the architecture of the microprocessor that the software will run on, the intricacies of the tools used to build and test the software, and a set of standards prescribed for the software's structure. In addition to this technical computer science knowledge, the successful development of avionics programs requires an understanding of numerous knowledge domains involved in the application. These domains include aeronautical engineering, aerial ordinance, flight navigation, electronic countermeasures, and probably an esoteric branch of mathematics that provides algorithms for specific avionic problems. Students cannot acquire all of this knowledge in college. Students with degrees in computer science are novices in an application domain when they first graduate. They will spend their first several years on a job acquiring knowledge of the application. Some of the substantial performance variation observed among software developers can be attributed to the speed and depth with which they acquire knowledge of the application domain.

Continuous technological change. Advances in technology affect the languages and tools that are used in developing systems. Aerospace environments are now faced with changing from computer languages such as Jovial and Fortran to a standardized language for all Department of Defense contracts, Ada. Although the language is available, many of the tools needed to develop and test Ada programs have yet to be developed. The support environment for Ada programming will become available in stages over the next several years. New developments in software design methods are also forcing software engineers to request different types of development tools. Thus, there is continuous change in software engineering technology that requires programmers to engage in continuous learning about the tools of their trade.

High percentage of work spent learning. Although the standard life cycle model for software development assumes stable requirements, the truth is that they are always changing, especially on larger projects that take longer to complete. The longer it takes to develop a system, the longer customers have to think of changes they would like. Further, customers often do not understand all the capabilities they want in the system at the outset. In this sense they are learning about the application along with the developers. Prototyping presents an alternate process model for developing software. In the early stages of a software-development process, developers often do not know enough about an application to anticipate the customer's preferences. Prototyping allows developers to enter a learning mode and determine through a partial system what the customer actually wants in a final product.

Developers may not be the only people to learn during development. For instance, the Aegis radar system used by the Navy was acquired in stages. At first the Navy thought they were merely acquiring a ship radar. Quickly they realized that Aegis could be the cornerstone of battle management for the ship. Next they realized that Aegis could be the center of battle management for a flotilla. Finally they realized that Aegis could handle battle management for an entire fleet. Thus, the requirements for Aegis kept expanding with each delivery cycle. The assumption of a stable specification is naive in software development because both the customers and the developers are learning about the system through building it.

Importance of rationale. After Christopher Wren designed St. Paul's Cathedral in London, generations of stone cutters had to carry out his original intentions or the edifice would come tumbling down. This same phenomenon occurs in software, except that you cannot see software start to fall if you violate the integrity of its design. Problems are obvious only much later in the process and are much more expensive to correct. Thus, progress can be illusory because rapid progress early in the design phase may come at an expensive price in redesign once problems in the original concept are discovered.

Figure 5.3 shows a model of how software development is frequently organized in large corporations. There is a series of teams, each of which passes an intellectual artifact to the next team. The process starts with a small group of people specifying the desired behavior for a system. They pass their concept on to a definition team that that does the initial design. The design is passed to a development team that produces, tests, and verifies the system. They pass the completed product on to the delivery team, which installs it and trains the customer. Finally, the maintenance team fixes operational defects and enhances the system based on new requirements.

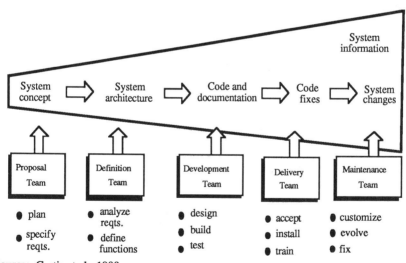

Source: Curtis et al., 1988.

Figure 5.3. Team communication by artifact.

Different forms of intellectual artifacts are passed between these teams. A crucial issue in system development is that the fundamental concept developed by the proposal and definition teams retain its integrity throughout its transformation by each succeeding group. The maintenance team needs to be able to determine why certain design decisions were made, and they frequently do not have updated design documents. Without understanding crucial constraints on design decisions, they may make changes that violate the integrity of the design and create new defects. In a strange way, there is an act of technology transfer being carried forward through this progression of transformations.

Heavy communication overhead. It is crucial to achieve consensus on the design and reach understanding on how established design decisions constrain later design decisions by other project members. Extensive communication is required to manage the constraints and integrity of the design. As the system gets larger these communication behaviors account for a larger proportion of how developers spend their time. In extremely large systems with constant requirement changes, there is a large constant overhead to ensure that responses to these changes are coordinated.

Figure 5.4 displays the communication nets programmers establish in order to acquire the information needed to perform their jobs. Some of these networks involve access to technical information about the application, the system architecture, available software components that can be reused, and the tools that are available for programming. The other communication nets involve management and organizational information. Different people are sought out for different forms of information, and maintaining the networks requires substantial time investment.

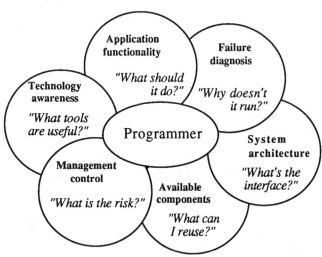

Source: Curtis et al., 1988.

Figure 5.4. A programmer's communication nets.

Hard-to-measure productivity, quality, and progress. How do you measure the quality or productivity of producing an intellectual artifact like a computer program or a research report? Jones (1986) chronicled the anomalies in software productivity measurement. So long as software development is treated as a craft rather than an engineering discipline, its performance will be treated as a judgment in the eye of the beholder rather than as an exercise in measuring a work product and the effort required to achieve it. Measurement of software development has been difficult because the work products are produced at many levels of abstraction. It is extremely difficult to define a rigorous method of representing the amount of product produced for a unit of time or cost. The number of computer instructions is the most frequent measure, although it has been roundly criticized for over a decade. Productivity and cost comparisons are usually performed within a single business environment because of the myriad factors that affect performance in one environment but not in another. Problems in establishing sound productivity and quality measurement has made it difficult to develop the financial and business arguments necessary for generating investment in technology transfer.

Technology Transfer in Software Organizations

> *The way the managers are getting trained is that the engineers are coming back [from software engineering courses] and are fighting to keep using some of the tools and techniques they've learned—and fighting against the managers to let them use them—and that's really how the managers are getting their experience.*
>
> Software engineer in the MCC Field Study (Curtis et al., 1988, p. 1274)

There are a range of excuses we hear for technology stagnation. Fundamentally, everybody says they are too busy getting current products out the door and do not have time to explore new technology, that the new technology is too risky, that the technology is not yet compatible with their product lines, etc. The fundamental issue is that developers and their managers are used to solving technical problems in accustomed ways. Inserting new technologies requires not only a change in their working behavior, but quite possibly in cognitive processes as well. Restructuring working behavior is frequently less difficult than restructuring knowledge to incorporate new problem-solving techniques.

In knowledge-intensive industries such as software development, the engines of production are people. We are not close to having expert systems that can build large systems, because expert systems are weak at integrating information across knowledge domains. Technologies that do not affect the intellectual processes of software developers are going to have little impact on productivity, quality, and costs. This is why data concerning the productivity impacts of new software tools and methods has been so disappointing.

Technologies that impact intellectual processes require extensive learning and practice. To integrate new technologies into the execution of their tasks, professionals must spend time not only learning the technologies and methods they embody, but also gaining experience by actually using them on sample problems in order to gain confidence that these tools can be used in actual practice. Practice on nontrivial problems is crucial before a skill can be truly mastered. However, few short-term training courses provide the level of practice required to allow professionals to use new technologies skillfully.

Technology transfer in a knowledge-intensive domain has to be modeled as a learning process. What does this imply about transferring software tools and practices into project use? Figure 5.5 presents what I observed while at ITT's Programming Technology Center concerning the variation in productivity as a function of how much a project used new tools and practices (Vosburgh, Curtis, Wolverton, Albert, Malec, Hoben, and Liu, 1984). If good practices or tools were not used, we uniformly observed low productivity. However, the fact that a new practice or tool was used did not guarantee high productivity. The large variation in productivity observed on projects that used new tools and methods suggests that their use may be helpful, but not sufficient for achieving high productivity. There is something more fundamental in achieving software development productivity than just using advanced technology and development practices.

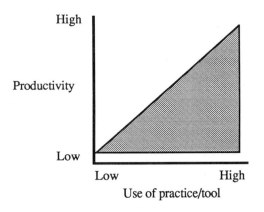

Figure 5.5. Variance in productivity as a function of technology use.

Let us consider an analysis of the processes that occur on software-development projects at several levels of analysis. The *layered behavioral model* presented in Figure 5.6 emphasizes factors that affect psychological, social, and organizational processes of system development, to understand how they subsequently affect productivity and quality. The layered behavioral model focuses on the behavior of those creating the artifact, rather than on the evolutionary behavior of the artifact through its developmental stages.

At the individual level, software development is analyzed as an intellectual task subject to the effects of cognitive and motivational processes. When the development task exceeds the capacity of a single software engineer, a team is convened and social processes interact with cognitive and motivational processes in performing technical work. In larger projects, several teams must integrate their work on different parts of the system, and interteam group dynamics are added on top of intrateam group dynamics. Projects must be aligned with company goals and are affected by corporate politics, culture, and procedures. Thus, a project's behavior must be interpreted within the context of its corporate environment. Interaction with other corporations either as co-contractors or as customers introduces external influences from the business milieu. These cumulative effects can be represented in the layered behavioral model. The size and structure of the project determines how much influence each layer has on the development process.

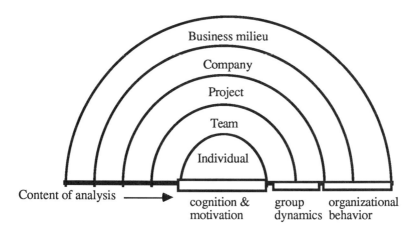

Figure 5.6. The layered behavioral model of software development.

The layered behavioral model is an abstraction for organizing the behavioral analysis of large software projects. It is not intended to replace traditional process models of software development, but rather to organize supplementary process analyses. This model is orthogonal to traditional process models by presenting a cross-section of the behavior on a project during any selected development phase.

Let us consider the mechanisms of technology transfer at these different levels of analysis. At the individual level, the technology transfer process frequently results from the efforts of a technology champion. In this model, the champion locates the technology and then takes responsibility for making it work and spreading the gospel. In our efforts to transfer technology from MCC's Software Technology Program, we have found the most successful model to be at the team level, where we establish a collaboration between our research team and a development team that is interested in applying the technology. In this model we work jointly with a program participant to tailor the technology to the idiosyncratic needs.

Technology such as computer-assisted software engineering (CASE) environments must be transferred at the project level to be effective. Such technologies must be used by all project members in order to have an impact. Transfer at this level requires management involvement and commitment to make an investment of this magnitude work. Transferring technology to achieve continuous growth requires a managed growth plan. Such a plan is only possible at the company level, because it requires commitments that outlast a single project. Implementing statistical quality control would require a company-level commitment, since the data acquired in such an effort are only meaningful if evaluated for steady quality growth over the course of several projects.

Some technology transfer occurs at the market level, because emerging standards force projects to acquire certain technologies. In the future most computer programs will have to work with window systems that work on top of operating systems, especially X-windows, because this technology is an emerging standard in the marketplace. Incompatibility with window systems will limit a product's sales potential. At the level of the polity, the government can be a force for transferring some technologies. For instance, the Department of Defense is requiring all of its contractors to use the Ada programming language. At each of the behavioral levels a different type of technology transfer mechanism is appropriate.

Three different patterns emerge for the productivity drivers at the three companies I have been with before MCC; Weyerhaeuser, General Electric, and the ITT Programming Technology Center (Figure 5.7). In older industries such as lumber production, applying more capital and labor has historically generated much of the productivity growth. In the 1950s Robert Solow (1957) realized that we could not account for productivity growth with just those factors and that technology development played a major role in many industries such as the manufacturing businesses in General Electric.

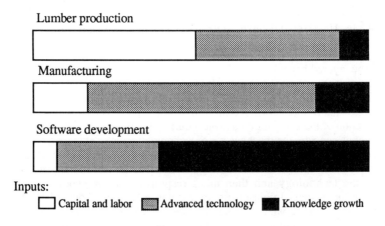

Figure 5.7. Sources of productivity growth in different industries.

However, in knowledge-intensive industries neither capital and labor nor advanced technology are accounting for very much of the productivity growth. The crucial new factor, especially in knowledge-intensive industries, is knowledge growth. The more knowledge the work force acquires, the more productivity is affected. The 20:1 differences among programmers are not strongly affected by the existing technology we have tried to install.

Figure 5.8 presents data developed by Sam Redwine and his colleagues a few years ago regarding the typical 18-year lag between the initial conceptualization of a technology or practice and its integration into popular use in software engineering. In the knowledge-intensive business of software development, this lag is partially the time required to perfect the technology and adapt it from the laboratory to actual industrial use. However, much of this lag is the time required to spread knowledge of how to use the technique across industry, with the accompanying experience base that the technique increases productivity or quality.

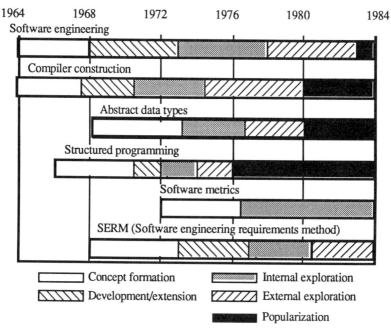

Source: Redwine et al., 1984.

Figure 5.8. Technology transfer lags in software engineering.

Future productivity growth in software development, and probably in many other knowledge-intensive businesses, will require technology that expedites knowledge growth across an entire team. Such technologies would include advanced communication facilities, information access and manipulation capabilities, and work technologies that speed the learning process. These are the technologies that will transfer the fastest in a knowledge-intensive business, because they are immediately perceived as solving the hard problems. They aid the learning process rather than require extensive learning for a minimal payoff. Thus, technology transfer in knowledge-intensive organizations is different and must be based on different drivers than it has been in other business domains.

References

Boehm, B.W. *Software Engineering Economics.* Englewood Cliffs, N.J.: Prentice-Hall, 1981.

Curtis, B., H. Krasner, and N. Iscoe. "A Field Study of the Software Development Process for Large Systems." *Communications of the ACM,* vol. 31, no. 11, 1988, pp. 1268–87.

Jones, T. C. *Programming Productivity.* New York: McGraw-Hill, 1986.

Redwine, S. T., L. G. Becker, A. B. Marmor-Squires, R. J. Martin, S. J. Nash, and W. E. Riddle. *DoD Related Software Technology Requirements, Practice, and Prospects for the Future.* IDA Paper P-1788, Arlington, Va.: Institute for Defense Analysis, 1984.

Solow, R. "Technical Change and the Aggregate Production Function," *Review of Economics and Statistics,* vol. 39, 1957, pp. 312–20.

Valett, J. D. and F. E. McGarry. "A Summary of Software Measurement Experiences in the Software Engineering Laboratory." *Journal of Systems and Software,* vol. 9, no. 2, 1989, pp. 137–48.

Vosburgh, J., B. Curtis, R. Wolverton, B. Albert, H. Malec, S. Hoben, and Y. Liu. "Productivity Factors and Programming Environments." *Proceedings of the Seventh International Conference on Software Engineering.* Washington, D.C.: IEEE Computer Society, 1984, pp. 143–52.

6

Science as a Competitive Weapon: Utilizing Alliances in Global Markets

Debra M. Amidon Rogers

Who would have dreamed of the radical changes occurring with the dawn of the unified European Community, the *glasnost* and *perestroika* of the Soviet Union, the psychological and physical removal of the great wall between the East and West, and the industrial advances of Asian and other underdeveloped countries—all leading to a new international economic order for which there is no blueprint. The transformations across countries and continents have significant implications for those corporations and institutions able to adapt to the dynamics of an evolving worldwide marketplace.

The Global Regime

Robert Reich (1987) in an article in *Atlantic Monthly* juxtaposed the two sides of the debate: "Techno-nationalism vs. Techno-globalization." In two words he was able to expose the myriad tradeoffs and tensions that exist across political, financial, and economic boundaries. Wise practitioners in the field were quick to realize that successful strategies were not simply "either/or" decisions. They were deliberate initiatives that combined and balanced both scenarios simultaneously to meet the diverse demands of the customer base. George Kozmetsky (1987) described it as effectively managing the paradox of cooperation and competition. Christopher Bartlett and Ghoshal Sumantra (1987) defined "the simultaneous demands for global efficiency, national responsiveness, and worldwide learning." Mel Horwitch (1990) used the term "simultaneity" (i.e., the "concurrent, purposeful positioning of seemingly diverse and contradictory aims and contexts"). This is the essence of global management.

All these visionaries seek integrative, systematic solutions resulting in a fusion of interests, represented in Figure 6.1 as total quality commitment. In some constructive way, the interdependent goals in industry, academe, and government can harness creativity across the sectors toward prosperous innovation.

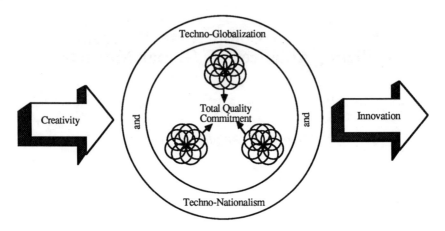

Figure 6.1. Toward paradigm integration.

In the October 1989, issue of *Scientific American*, Reich turns toward the exploitation of science and R&D as the "quiet path to technological preeminence." Not only does he argue for the international flow of technology, but he articulates the six steps necessary for U.S. corporations to regain their competitive edge. They all focus on the transfer process itself.

Step #1: Organize for systematic global scanning as an integral part of business strategy.

Step #2: Improve the assimilation of public-funded research by linking it directly to commercial production.

Step #3: Integrate internal R&D processes with the production and marketing functions of the company.

Step #4: Participate in the early adoption of industry-wide standards to speed commercial acceptance.

Step #5: Develop a sophisticated work force eager to adapt a broad range of new technologies.

Step #6: Provide a foundation for basic education and, in so doing, the ability to learn.

In these specific guidelines, Reich (1989) contrasts the differences between Japanese and U.S. management with the subtle recommendation of how U.S. firms might capitalize on their own strengths. The outline is clear and simply stated, but the implementation is very complex.

The "One-Minute Innovator"

In 1982, Kenneth Blanchard and Spencer Johnson authored *The One Minute Manager*, which was intended to offer some key, simple techniques to influence management practice. It suggested quick ways to increase productivity—on both individual and organizational levels. Through an allegory, the authors synthesized crisp messages based on the wisdom gleaned from studies that some might describe as cross-disciplinary (i.e., medicine, behavioral sciences, and management).

Amid the complexity of the process of technological innovation and the lessons learned from recent collaborative activities and studies, I have come to the realization that what we really seek is a "just-in-time research" environment (or research "just in time," if you prefer). One can hardly argue with the seven basic tenets of "total quality" environments: commitment, simplicity, flexibility, timeliness, cost-effectiveness, people-intensiveness, and continuous process improvement. In this scenario, the technological breakthrough that occurs within the university research laboratory reaches the advanced development workbench engineer of the sponsoring corporation, in "real-time" measures (i.e., when there is a "need to know"). Likewise, the insight of the industrial engineer can provide timely influence on the research progress of the academic scientist. This is the symbiotic relationship we seek in which there is an optimal "knowledge transfer" infrastructure.

To this end, I offer four principles as simple reminders of how we might strengthen the entire process of innovation.

Principle #1: Manage the Process of Technology Transfer

As mentioned earlier, the international competitive environment mandates that we no longer leave the technology transfer process to serendipity. Not that informal, unexpected connections are not the best. In many cases, they are. But we must be far more fastidious in systematizing how we manage the knowledge flow into and throughout our U.S. corporations. Designating people responsible for technology transfer is a beginning.

Principle #2: Build Quality into the Front End of Research Relationships

Reams have been written on the obstacles to successful technology transfer. Let us focus instead on what makes it work. "Predictors of success" could offer a step toward that insight. Build these quality factors into the early stages of research programs, rather than treating them as afterthoughts. Let's view academe, industry, and government as three integral partners, each with its own paradigm but in agreement that increased quality is the goal we collectively seek.

Principle #3: Practice the Art of Continuous Process Improvement

Let's stop reinventing the wheel. There is much to be gained from the lessons of previous programs. Too often organizations, agencies, and people themselves are building the equivalent of their own empires, based on the need to structure better academic-industry research relationships. There is now a need for synergy and cooperative, versus competitive, efforts. The Japanese are experts on building on the past and on continuously employing new approaches in refining processes. Let's focus on what we already do well and what we might do better.

Principle #4: Optimize through Alliances

Given the dynamics of a worldwide marketplace and what has been foreseen as a hyper-competitive environment, it is difficult to comprehend a corporation that can successfully go it alone. There is strength in some cross-boundary interconnections at multiple levels—internal to the firm with cross-functional learning, as well as external—cross-disciplinary, cross-industry, up and down the value chain, and across geographic boundaries. Only through skillfully partnering can corporations practice collaborative strategies that maximize a worldwide competitive advantage. The rate of technological change and the intensity of market dynamics require novel business management approaches.

Such joint ventures, especially across international boundaries, are not simple. As described by Karen Hladik and Lawrence Linden (1989) in a recent issue of *Research Technology Management,* one must enter such alliances in full awareness of the pitfalls involved:

- Risks of sharing proprietary know-how
- Issues of control and product design
- Dissimilarities between potential partners
- Integration and communication with the rest of the parent company
- Antitrust regulations and patent protection.

But a *Fortune* article by Jeremy Main (1989) also suggests the need to find allies in order to creatively position a company. Companies need to address multiple dimensions in order to maintain future competitiveness. Paraphrased, these include: creating a tri-geographic presence, developing products for the world, establishing "product-line" versus "geographic" profit centers, "globalizing," overcoming the NIH ("not invented here") syndrome, opening senior ranks to foreign employees, doing whatever seems best (wherever best), and finding allies. Throughout the literature, then, there seems to be a unified message and a realization that cooperative activities are the way to go.

Technology Transfer: Lessons Learned

In 1987, a round table was convened to focus on the issue of technology transfer within U.S. research consortia, the proceedings for which are entitled "Managing the Knowledge Asset into the 21st Century" (see Rogers and Dimanceascu, 1988). Although limited by the simplistic, linear model of the technology transfer continuum, the meeting did provide insight into the innovative management agenda that requires a new, concurrent set of activities, capitalizes on the continuous nature of human interactions, creates a "real-time" process, and calls for new organizational tools that are "market-pull" versus "technology-push" driven.

For all the analysis, however, I often wonder how much closer we are to truly understanding the intricacies of the processes required for twenty-first century competitiveness. There is a book entitled *The Goal: A Process of Ongoing Improvement* by Eliyahu Goldratt and Jeff Cox (1986) that has provided a breakthrough in my own thinking. Given the realities of a transformed worldwide networked technology communication base and the sophistication of the consumer, "new age" organizations will have to be structured in radically different ways.

The transfer continuum is converted into a lotus flower to depict the integration of functions required in contemporary competitive business strategy. Most U.S. corporations are certainly not structured to capitalize on the interdependencies represented within the petals of the flower (e.g., where research meets marketing, where manufacturing meets services, etc.). Previous patterns of management—even in multidimensional matrix organizations such as Digital Equipment Corporation—do not lend themselves to resource allocation practices that are not functionally driven. Interestingly enough, in the corporation as a whole, information technology—or the use thereof—seems to transcend traditional functional relationships. The "peer-to-peer" networking system is a communication tool that encourages the cross-organizational learning required in the competitive knowledge network of the future, and it is inherently global in nature.

It is not enough, however, to create such an open-network environment exclusively within the firm. There are numerous interdependencies with external institutions, professional organizations, government agencies, industry partners, research consortia, and other joint ventures that require appropriate leveraged interfaces throughout the firm. Concurrently, to add to the complexity, such interrelationships must be created worldwide to catalyze global competence. Such is the agenda for many of the alliances in which Digital participates. These include Alvey, Esprit, Sprite, etc. In fact, Digital's internal research operation has global outposts in Canada, Japan, Australia, and throughout Europe.

This fusion of knowledge is what Sheridan Tatsuno (1988) describes as the difference between Western and Japanese styles of management:

> Western creativity is based on the notion of individual freedom and expression. It is the nuclear fission in which individual atoms produce energy; by contrast, Japanese creativity is more like nuclear fusion, in which particles must join together to create a reaction.

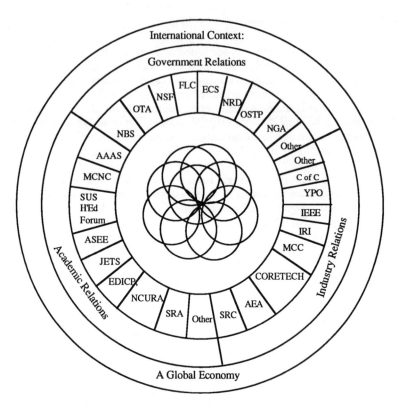

Figure 6.2. The global environment.

It is incumbent, then, upon modern innovation managers to be able to decipher simple, targeted strategies amid the complexity and to harmonize the creative tensions inherent in every interconnection at which something must be delivered and/or received.

Crafting a Global Innovation Strategy

These thoughts became the heart of my masters thesis at the Massachusetts Institute of Technology (Rogers, 1990). The thesis research provided opportunities that helped shape a new paradigm for innovation management. They allowed me

- to return to the "art of strategy," based on Sun Tzu's *Art of War*
- to draft two basic propositions essential for progressive management
- to study in depth three joint ventures with cross-geographic implications

- to expand the framework of the lotus flower as a way of conceptualizing novel structures, resource allocation practices, and communication and business-planning processes.

The MIT experience legitimized the role of "strategist"—not that of a long-range planner or an organization-development consultant. The strategist takes a holistic view of the entire system with its multiple interconnections of parts. It also requires an understanding of socioeconomic dynamics and how they might be harnessed for bottom-line profitability and customer satisfaction. Here are six brief messages from a book entitled *The Art of Strategy* (Wing, 1988) that guided much of the analysis revealed in the thesis. It is taken from Wing's translation of Sun Tzu's military treatise. His philosophy is one of coherence, flexibility, and tactical positioning—not too dissimilar from the challenges afforded innovation managers today.

- Study of strategy cannot be neglected
- "Tao of Paradox" or control through confusion
- Overcoming adversity through a combination of advantages and disadvantages
- "Five Strategic Arts" from measurement through balance toward triumph
- Winning through "systematic positioning"
- Engaging the total system in a "plan of attack."

If practiced daily throughout the business routine, these are the types of basic precepts that can incrementally change one's perspective of the business. Ultimately, they appear to provide a path to the transformation required for business managers to skillfully play the "video game" of the new global industrialized complex.

Second, two propositions form a foundation for analysis of the effectiveness of a function within a firm, a company within an industry, and/or companies in some form of joint partnership.

Proposition #1: A company must employ an innovation strategy that integrates functions such that the system enhances technological breakthroughs to reach the marketplace in real time.

In other words, as soon as a research insight or breakthrough is reached in the mind of a researcher (whether inside the company, at a university, at a government laboratory, at a research consortium, or in a joint venture—and wherever it may be worldwide), a product is created, manufactured, marketed, and ready to be received and serviced instantaneously. The cycle time in the process of innovation is reduced to zero. In reality, such measures are hardly achievable; but when such an ambitious goal is set as a target, it does prompt one to structure and allocate resources, both financial and technical, very differently. It also puts into place the need for radically different business planning and evaluation processes.

Proposition #2: This strategy to be competitive must be driven globally with a balance of effort along intraorganizational, interorganizational, and transnational interfaces.

This does not mean that such activities cannot occur along only one axis (e.g., intraorganizationally). Many such initiatives do, and some are successful. But to be an optimal strategy (i.e., one that creates the greatest amount of leverage for given resources), it must be driven along the three axes simultaneously, as depicted in Figure 6.3.

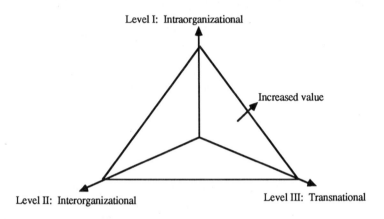

Figure 6.3. Balance of economic perspectives.

In the alliances researched for my thesis, the contrast of a truly global competitive venture versus an incremental added-value partnership became quite obvious. The three alliances studied were the Digital/MIPS Co. Alliance from the U.S. perspective; the BiiN:Siemans/Intel Alliance from the European perspective; and the HN Bull/NEC Alliance from the Japan perspective.

Each alliance could be plotted according to the dimensions of strategy designed by Michael Porter (1987). They differed in their configuration of activities from high to low and the coordination of activities ranging from geographically concentrated to geographically dispersed. One could easily argue the relative position of each depending on the value activity defined. That is not as important as being able to differentiate—even in some simplistic way—the differences between the alliances. Each venture was profiled including the ideologies fundamental to the geographic perspective. The second dimension of analysis included three economic levels:

- Intraorganizational (microeconomic or firm specific)
- Interorganizational (mesoeconomic or industry-wide, national or domestic)
- Transnational (macroeconomic, global or international).

The third dimension of analysis defined three strategy elements:

- Structure: organization configuration (e.g., centralized versus decentralized, shallow versus deep), project management (e.g., team versus hierarchical, vertical versus horizontal integration, national versus international infrastructure)
- Resource allocation: financial, technical, and human resource (e.g., long-term versus short-term, quality of expertise, incentives, rewards, etc.)
- Process: spans communication, planning, and decision making (e.g., electronic or programmatic, feed-forward intelligence systems, bottom-up versus top-down or horizontal, incremental versus quantum, explicit versus implicit, traditional versus avant-garde).

All three dimensions are reflected in the cube in Figure 6.4 which depicts the framework used for analysis.

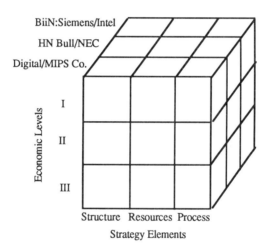

Figure 6.4. Thesis framework.

The findings reaffirmed the phenomenon of simultaneity defined earlier. It is the balancing of the seeming opposites of the continuum that strengthens the adaptability and flexibility required in the system to withstand the competitive dynamics to come. This is true when any two firms (or functions, for that matter) meet in a joint venture for mutual gain—especially when the dynamics of a large firm and a small firm are at play.

Figure 6.5 by John Friar and Mel Horwitch (1986) best depicts the creative tensions inherent in such partnerships. There are limits and boundaries within which each must work. It is likely that in the resolution of these tensions, one finds the value creation of cooperative ventures. It was clear that focusing on similarities rather than differences also enhanced the partnership. It was the cross-boundary interfaces, where risk could be shared, that legitimized many of these relationships. And in all cases, it was the opportunity to configure something differently—the experiment, if you will—that increased the learning of all partners involved, regardless of the success of the venture in and of itself.

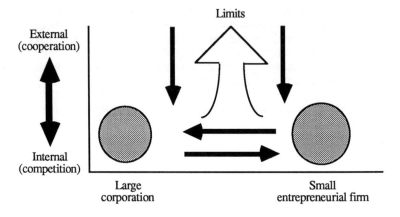

Figure 6.5. The Horwitch model of joint ventures.

Finally, there is the new conceptualization of the innovation management process that can be depicted with three tiers of the lotus flower: intra-organizational, interorganizational, and transnational. At the heart of the multiplicity of layers lies the common core quality to the customer (which, by the way, is simultaneously profitable to the corporation).

The "Lotus Flower" Framework

Many of the modern frameworks for analyzing technology strategy, innovation management, strategic alliances, and national economic enterprises are becoming so complex that we are unable to capture the essence—the simplicity in design. A return to nature with the creation of a lotus flower provides a graphic representation of the interconnections and interdependencies that must evolve within a corporation's management structure, within an industry or domestic strategy, and across national boundaries.

For the purpose of this description, we will define the three levels as intra-organizational, interorganizational, and transnational. It is apparent that these levels are not independent; in fact they are anything but mutually exclusive. The idea is, however, to define them as three separate entities for the purpose of analysis.

Level I: "Intraorganizational"
(i.e., microeconomic or firm-specific)

Each company is comprised of separate functional units or groups within groups. These have been defined in many ways, most notably through Porter's (1987) "value-chain." They represent the activities within the firm that are necessary to transfer or transform R&D discoveries into emerging designs: manufactured, marketed, and sold to the customer, usually with services included. However, to date these illustrations are linear in appearance. Consider, instead, a series of concentric circles, each intersecting with one another in some undefined way. Admittedly these circles would not necessarily be equal depending on the criteria of measurements used, but the concept of intersecting boundaries within functions is useful. The resultant diagram below represents the first tier of the lotus flower.

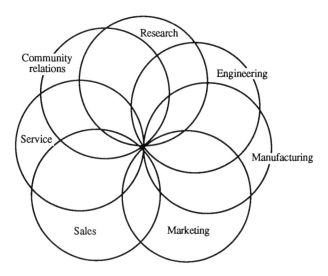

Figure 6.6. First tier of the lotus flower at the intraorganizational or microeconomic level.

Note that with this perspective we can begin to explore the opportunities created by functional integration. Defining the interrelationships that could be created between services and manufacturing or R&D and marketing, for example, suggests new alternatives for structure alignment and program activities. The value of competition within the firm for budget and technical resources not withstanding, this model depicts some new interfaces that might better create competitive advantages. There is research under way to study the implications of streamlining and integrating functional activities for efficiency and effectiveness. Professor Gabriel Bitran, MIT department chairman of management science, has defined in a lecture the problem as "producing a product that is more technologically innovative, on time, and at a price acceptable to the customer." It is that simple and that complex at the same time. He continues, "Service operations and manufacturing have more in common than many people think. One can realize after a while that it is not important to stress differences, it's important to stress similarities." This is the type of thinking required across all functions in the corporations to transform theory into viable competitive enterprises.

Level II: "Interorganizational" (i.e., mesoeconomic or industry-wide, national or domestic)

It is becoming increasingly more important for corporations to establish vertical linkages both upstream and downstream in the "value-chain." This may not—most likely does not, given current economic conditions—mean vertical integration within a corporation. Admittedly, such acquisition on either end of the chain can provide some economies of scope advantages. For the most part, forming solid working partnerships with key suppliers and distributors will minimize risk. But these relationships are dynamic in nature, evolve over a period of time, and require that people separated by distance, time, culture, and language may have difficulty in forging alliances. This has certainly been the case with the cross-sector adversarial relationships within the United States between government, academicians, labor, and industry. Some balance of power mechanisms (e.g., antitrust laws) have become counterproductive in this global arena. Instead, perhaps we can envision the second tier of the lotus flower building upon the first. See Figure 6.7.

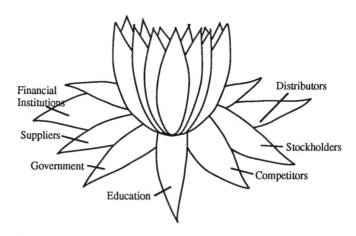

Figure 6.7. Second tier of the lotus flower at the interorganizational level.

It is not sufficient for a corporation to focus solely on internal functional integration when there are external interdependencies that must be equally managed to ensure that the firm is able to position itself within its desired market. Interfaces might need to be strengthened with the financial community, suppliers, government entities (including federal and national laboratories), academic institutions at all levels, stockholders, distributors, and even competitors within the industry. These entities, in and of themselves, do interact with one another as well. Much does happen serendipitously. However, a great deal of the interaction might well be more carefully structured to ensure the best utilization of scarce corporate resources. It is also the only way a national economic enterprise can be formulated.

Japan has created such an infrastructure to surge to technological supremacy with the Ministry of International Trade and Industry (MITI). Gorbachev attempted to transform the Soviet Union with *perestroika*. The European Community is moving toward an ambitious plan to create a unified Europe by 1992. Examples of industrial integration within the United States include the creation of the Microelectronic and Computer Corporation (MCC) and the Semiconductor Manufacturing Technology Initiative (SEMATECH), a consortium of suppliers, manufacturers, and government and academic resources organized to address the U.S. semiconductor dependency problem.

Level III: "Transnational"
(i.e., macroeconomic, global or international)

Advancements in information technology and other technological innovations have created a worldwide laboratory and marketplace within which corporations, large and small, compete on competitive and comparative advantage bases. This is the arena in which we see the national economic programs referred to in Level II operating. It is also where we discover the transitions, within corporations, from domestic to multinational and eventually international and transnational strategies.

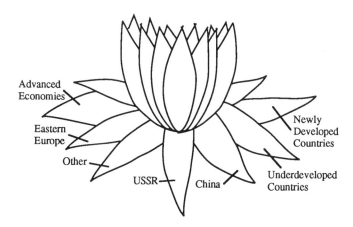

Figure 6.8. Third tier of the lotus flower at the transnational level.

52 • *Technology Transfer in Consortia and Strategic Alliances*

The world is comprised of numerous players representing individual countries on the one hand or groupings represented by recent taxonomies: North America, Western Europe, Eastern Europe, Japan, Asia, Russia, Newly Industrialized Countries (NICs), and Underdeveloped Countries (UDC). The implication of this tier is that there is value in increasing the size of the potential marketplace as well as linking with worldwide partners for various functional aspects of the "value-chain." This is where global sourcing and adaptive R&D strategies can be most effective. In addition, capitalizing on the strengths of global partners can provide early entry into some countries, and automatic market share.

Selecting partners and managing joint ventures usually end up being far more complex than originally intended. But that is the nature of the worldwide economic enterprise. Central to the theme of this tier—although not limited to it—is the concept defined by Bartlett and Sumantra (1987) of three simultaneous flows to be analyzed: (1) flow of parts, components, and finished goods; (2) flow of funds, skills, and other scarce resources; and (3) flow of intelligence, ideas, and knowledge. Once again we recognize the trade-off tensions created by the simultaneous objectives, but also the opportunities that can be created by a worldwide infrastructure in which a given corporation might do business.

This lotus flower analogy is not complete without one final reference to the core of its being—the customer orientation. Not only does the complete blossom represent a coherent concept of global innovation strategy, but its central pistil can be depicted as "quality" for the customer and the firm. With such a common focus, all efforts at each tier can be appropriately synchronized. See Figure 6.9.

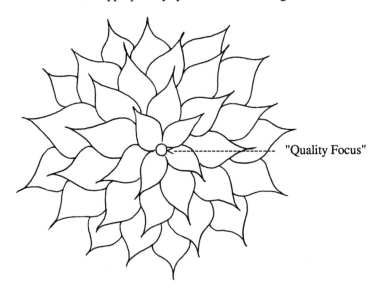

Figure 6.9. The core of the lotus flower: the customer.

As parts of the "innovation lotus flower" overlap, the differences become more apparent. But it is precisely within the boundaries of these overlapping interfaces that similarities may be discovered and leveraged. This is the nature of the modern business of "value-creation opportunities" as it relates to cross-functional integration within a firm, cross-sector integration within a nation, and cross-cultural integration in the global infrastructure. Only a comprehensive competitiveness policy that nurtures such integrated business systems will enable corporations, industries, and indeed nations to survive in the decades to come.

Crafting Change

Marketing this type of thinking operational is hard work because of the nature of the boundaries crossed and the resulting creative tensions. But the old adage says, "If it's difficult to do, it's likely worth doing . . . at least in business." And so, within Digital there are multiple experiments, pilots, vision formations, and the like. One example is the new CIM (i.e., Complex Interrelationships), which integrates Digital with suppliers, customers, and other significant external parties. Within the services organization there are major Enterprise Integration Systems (EIS) demonstration sites. Worldwide, Digital has established a dozen Digital Competency Centers (DCCs) that focus on a particular industry (e.g., finance, manufacturing, telecommunications, etc.), but provide resources and expertise worldwide.

The framework in Figure 6.10 outlines the necessary elements to be fused into a single strategy: international vision, technology strategy, market strategy, and strategic human resources. Concurrently, direct attention needs to be paid to the competitive analysis "feed-forward" systems, communication processes, tri-level process integration, and optimization of alliances. To create such a strategy requires systematic executive management intervention.

Digital's problem is not a lack of vision. We have multiple, bold visions that now, within the corporation, need to be fused into a coherent strategy; simple and profound enough to fire the imagination of our 120,000+ employees worldwide.

Conclusion

I am amazed at the transformation that has occurred not only within Digital but within the worldwide economic enterprise as a whole. Science will continue to drive the innovation process, but with a new sense of manufacturability, marketability, and serviceability in mind. Science can be used as a formidable competitive weapon, but only if the system—broadly defined—is ready to integrate and assimilate it into commercializable product.

Creation of the global innovation enterprise is necessary if we are to take full advantage of our intellectual capability to create wealth and an improved standard of living. This means a new management agenda that creates real-time innovation, establishes a global competitive balance across all three economic levels, and internalizes the true art of strategy.

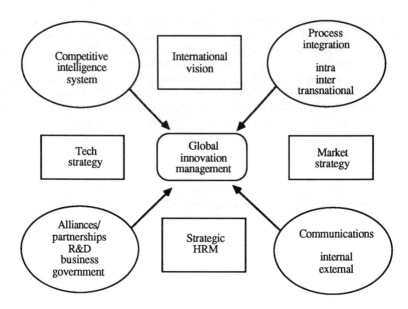

Figure 6.10. New strategy elements for global innovation management.

References

Bartlett, Christopher and Sumantra Ghoshal. "Managing Across Borders: New Strategic Requirements," *Sloan Management Review*, Summer 1987.

Blanchard, Kenneth and Spencer Johnson. *The One Minute Manager*. New York: William Morrow, 1982.

Friar, John and Mel Horwitch. "The Emergence of Technology Strategy," *Technology in the Modern Corporation*. New York: Pergamon Press, 1986.

Goldratt, Eliyahu M. and Jeff Cox. *The Goal: A Process of Ongoing Improvement*, 2nd ed. Croton-on-Hudson, New York: North River Press, 1986.

Hladik, Karen J. and Lawrence H. Linden. "Is an International Joint Venture in R&D For You?" *Research/Technology Management*, July–August 1989.

Horwitch, Mel. *Post-Modern Management: Its Emergence and Meaning for Strategy*. New York: The Free Press, 1990.

Kozmetsky, George. 1987 Annual Report, IC^2 Institute. The University of Texas at Austin.

Main, Jeremy. "How to Go Global—And Why," *Fortune*, August 28, 1989.

Porter, Michael. "Changing Patterns of International Competition," *The Competitive Challenge: Strategies for Industrial Innovation and Renewal*. Boston: Ballinger Publishing, 1987.

Reich, Robert B. "The Rise of Techno-Nationalism," *Atlantic Monthly*, May 1987.

Reich, Robert B. "The Quiet Path to Technological Preeminence," *Scientific American*, vol. 261. no. 4, October 1989.

Rogers, Debra M. and Dan Dimancescu. "Managing the Knowledge Asset into the 21st Century: Focus on Research Consortia," Proceedings of a Round Table, Technology and Strategy Group, 1988.

Rogers Debra M. *Global Innovation Strategy: Creating 'Value-Added.* Masters thesis, Massachusetts Institute of Technology, published by IC^2 Institute, The University of Texas at Austin, 1990.

Tatsuno, Sheridan. "Japan's New Challenge: Shifting to Creativity," *Dataquest Special Report,* October 1988.

Wing, R. L. *The Art of Strategy: A New Translation of Sun Tzu's Classic 'Art of War.'* New York: Doubleday, 1988.

7

Technology Transfer in a Diverse Corporate Environment

Wendell M. Fields

This chapter describes the diversity and the problems facing those involved in the process of technology transfer at Hewlett-Packard (HP), a company with a strong corporate culture. It presents a model for transferring technology that has been in place at HP for the last few years, along with descriptions of some of the programs that are being used to implement the model. Hewlett-Packard, like most large companies, transfers technology to a number of functional areas such as personnel, marketing, manufacturing, customer support, field service, and more. However, this chapter will focus on the functional area of research and development.

Corporate Culture and Diversity within Hewlett-Packard

Culture is defined as "the ideas, customs, skills, arts, etc. of a given people in a given period." Corporate culture is, simply stated, the way things are done. At Hewlett-Packard, despite its strong corporate culture, there is great diversity in the way things are done.

HP is a geographically diverse company with R&D labs throughout the United States, Europe, and the Pacific Rim. The varied geographic locations of HP employees creates problems associated with communication, time differences, and centralized training of engineers. In addition, each of the various locations is run with considerable autonomy, like a separate company with individual needs, skill requirements, and problems. HP produces over 10,000 products ranging from application software and operating systems to measurement/test instruments and scientific workstations. This creates a different business environment (different technologies, customers, and competitors) in each individual entity. Furthermore, the decentralized nature of HP has led to the problem known as the "not invented here" (NIH) syndrome, where individual entities will tend to be slow to adapt a technology that was not initiated locally. When technology comes from another entity within HP or from outside of HP it is sometimes not quickly accepted. NIH is another element of the corporate culture that affects technology transfer.

In order to support a wide range of products, a variety of engineering disciplines are supported at HP. In R&D, the needs of electrical engineers, mechanical engineers, human factors engineers, and software engineers must be met. To compound the problems associated with the diversity of engineering disciplines, each discipline has practitioners who specialize. For instance, in the area of software engineering HP has employees who work on applications software, operating systems, embedded systems, data programming, process programming, heuristic programming and more.

While differences in geographic location, engineering disciplines, NIH, and the business environments of HP entities contribute to the challenge of technology transfer, another major influence is part of the corporate value system itself. From its earliest days Hewlett-Packard has placed a high value on the importance of the individual employee. This value is explicitly stated in the set of corporate values known as "The HP Way" and exemplified by such programs as flexible hours, profit sharing, service awards, "informal get-togethers," and many others. It is this very emphasis on individualism that often hinders change through technology transfer.

Individual projects are empowered through management by objective (a valued company management technique) to do things "their own way." The concept of management by objective at Hewlett-Packard is defined in HP's *Corporate Policies and Guidelines Manual* as follows:

> Insofar as possible, each individual at each level in the organization should make his or her own plans to achieve company objectives and goals. After receiving managerial approval, each individual should be given a wide degree of freedom to work within the limitations imposed by these plans and by our general corporate policies.

This freedom is further reflected by The HP Way, which states, "We create a work environment which supports the diversity of our people and their ideas."

As HP has grown in size and complexity, the importance of company-wide standards for how products are designed is becoming clear. But, in a company that values management by objective, standards are a difficult thing to introduce. HP engineers do not mind being told what needs to be done, what quality level is expected, and the time constraints involved. However, they do mind being told how to do it.

Another change in HP's corporate culture has become apparent over the last few years. In the past an engineer could expect to follow a career path that moved from being a contributing engineer into the ranks of management. This is generally no longer the case. Due to organizational down-sizing as well as an older work force, there are fewer management positions available and it is more likely an engineer will remain an engineer much longer than in the past. Figure 7.1 identifies the variety of levels in the R&D engineering ranks at Hewlett-Packard. While they are not recognized by HP's personnel department, I believe that they represent observed stages in an engineer's development at HP.

When engineers enter HP they possess the technical skills and training needed to begin being an individual contributor to a project. As they gain more knowledge of enabling methodologies, the contribution level is more oriented toward components of the total life cycle. A firm understanding of the tools and how to measure the process distinguishes the next level engineer. Traditionally, engineers at this level move into the ranks of management. However, as engineers stay engineers longer, we are seeing other levels of engineering expertise. Engineers are now specializing in the process and contributing to its improvement, while a small percentage of our engineers are acquiring a high level of global expertise, which is contributing to the strategic impact of technology to Hewlett-Packard's business. The challenge is to allow engineers to develop through these various levels of engineering.

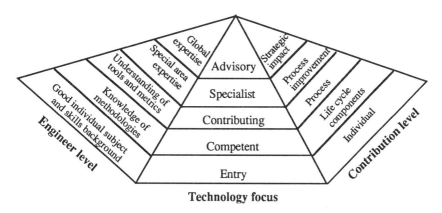

Figure 7.1. Pyramid of individual development.

A Model for Technology Transfer

In view of the issues and HP's corporate diversity and culture, we have found that a multifaceted approach to technology transfer is most successful as depicted in Figure 7.2. For a technology to be accepted at Hewlett-Packard there must be five factors in place. While change can begin with any of the five factors, it is more likely to be successful and permanent if all factors are at work.

It is important that a technology be considered a meaningful, value-added change that will positively impact the business of the company. This is accomplished through high-level endorsement, which in this case means leaders in an organization that are accountable for elements of the process that is to be changed. One means of informing higher-level managers of the importance of a technology to their business has been executive briefings. These are half-day conferences in which the aspects of the technology are presented less from a technical view and more from a view of how the technology can improve the bottom-line profit of the business. Another method that HP has used to introduce

60 • *Technology Transfer in Consortia and Strategic Alliances*

technology and gain high-level endorsement has been through management overview courses. This type of course has been successful in educating the managers of HP in such technologies as artificial intelligence and object-oriented software programming methods.

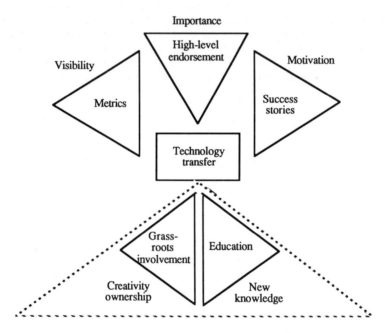

Figure 7.2. Hewlett-Packard's multifaceted approach to technology transfer.

Visibility happens when measurable outcomes of a technology's use or a change in process are identified. Hewlett-Packard has taken an approach of measuring the design process in individual corporate entities. Thus far, the measurement effort has concentrated on software development but plans are in place to include all R&D disciplines. The results are compared to industry standards in nine areas that include programming environment, project management, staff variables, noncoding tools, physical environment, measurement, defect removal and prevention, and methodologies.

The process for gathering the information involves data gathering teams that visit individual R&D labs. During the visits a team interviews all levels of technical employees in the lab: engineers, project managers, section managers, and lab managers. Questions are aimed at measuring the processes and comparing them to industry standards. Before a visit is complete, the visiting team presents the findings of the interviews. Descriptive statistics are presented to the entire R&D lab showing where they rank in relation to industry standards in the nine areas. Recommendations are made that are aimed at improving the process in areas that are weak. These recommendations often include the introduction of a new technology that will help improve the process. In addition to measuring the use of technology inside HP, data are also collected on the use of technology by HP's competitors in selected businesses. In an environment where top-level managers want to be world-class competitors, it is helpful for HP managers to see how other companies are using technology.

When a technology has been applied successfully to a project in HP it is important to tell the success story. Success stories provide motivation for other projects to try the technology. HP has found a number of ways to tell its engineering community of successful applications of technology. One way is the use of videotape. HP has produced videotapes that document the successful use of technology by a number of projects. The tapes, which are distributed at no cost, have proved to be a strong motivator for HP entities to try a new technology or process change.

In addition to videotapes, success stories are told through publications. Every other month, a publication is distributed to all HP R&D engineers featuring stories of how a project successfully used a new technology. Normally written by a member of the successful project team, the stories include names and numbers of people to contact. Success stories are also shared through our training and education programs, and in particular, seminars, expert forums, and conferences.

New knowledge, through education programs, helps to generate grass-roots involvement in a technology. Grass-roots involvement creates ownership in the technology by involving people in identifying and solving problems and defining measures of success. This creates "champions" who will provide bottom-up support that leads to effective technology transfer. Education is considered a cornerstone for the improvement of all processes and is therefore a basic organizational value held at HP. The HP Way states, "HP people should personally accept responsibility and be encouraged to upgrade their skills and capabilities through ongoing training and development."

HP's value of education allows the transfer of technology through a number of effective educational programs. One such program is the leading-edge seminar series, a series of one-day conferences that bring together engineers for the purpose of learning about leading-edge technologies that will be important to their business within a two- to five-year time period. The seminar is presented in two formats. In the first, engineers gather at one location to participate. Because of HP's geographic diversity, in the second format the seminar is telebroadcast to HP sites.

Leading-edge seminars consist of speakers from both within and outside HP. Engineers and managers share their research or experiences with a technology. Outside speakers, who are recognized experts in the technology, contribute an important, credible perspective. Time is allowed for questions and for the informal exchange of ideas by the attending engineers. While the teleconference version of the seminar allows only limited exchange of ideas between attending engineers, it does allow for questions to be asked of the speakers and has proved to be a cost-effective way to present information.

The technology topics for each seminar are determined through various means. The technology for transfer may come from the HP technology councils, through executive demand, through engineering demand, or through research conducted by members of the technology transfer staff working in cooperation with engineers from throughout HP as well as academia.

These seminars are more than informational; they provide an opportunity for engineers to hear from their colleagues so that they may understand how a technology has been successfully implemented in other HP locations. In many ways these seminars address other facets of the model for technology transfer. The talks motivate, educate, and generate grass-roots involvement in the technology. Leading-edge seminars have been presented on a number of technologies, including artificial intelligence in manufacturing, superconductivity, neural networks, and cooperative computing environments.

Expert Forums

Expert forums are another educational activity designed to assist in technology transfer. The model of an expert forum, shown in Figure 7.3, consists of four sections that take place over a two-day time period. One of the main goals of the expert forum is to bring together practitioners of a particular technology or methodology. This differs from a leading-edge seminar in that the audience is more focused and actually practices on the technology. This simple idea is taken from the user group concept. We feel that in a large company such as Hewlett-Packard, it is important to have periodic gatherings of engineers with common interests to fuel their professional growth as well as their channels of communication. The expert forum series fills this need.

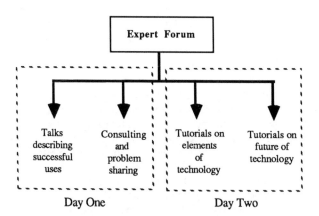

Figure 7.3. Hewlett-Packard's expert forums to assist in technology transfer.

The first section in the expert forum features case study presentations by internal HP people describing the use of the particular technology or aspects of the technology in their project. These experiential discussions, often referred to as "war stories," act as a catalyst to motivate discussion among members of the forum.

The focus of these presentations is on management issues in addition to specific aspects of the technology. During these discussions, many issues are raised and addressed. Much of the folklore surrounding the engineering process is discussed. This allows engineers to share ideas that cannot be learned in a classroom or read in a book.

Section two, consulting and problem sharing, allows novice users of a technology or methodology to take advantage of the collective wisdom of the forum members by presenting their problem and receiving real-time consultation. For instance, if members of a project are considering using a new technology, they can get valuable feedback from the members of the forum on the pros and cons of the technology as applied to their project. Following the presentations are technical discussions beneficial not only to the presenters but also to the members of the forum. This gives all of the engineers in attendance an opportunity to see how others would apply the technology/methodology to the presented problem and perhaps see alternate methods from which they would benefit in their own application of the technology.

The first two sections of the expert forum have been most successful at HP when an active audience facilitates the discussion. When the raised issues get addressed the audience is encouraged to actively participate. In the expert forum, model papers are not simply presented with a brief question-and-answer session. Instead, a brief overview is presented and much interactive discussion takes place. The third and fourth sections of the expert forum feature tutorials designed to expand the range of expertise and knowledge of the expert forum members.

The third section of the expert forum focuses on tutorials that teach creative and current best practices of the technology/methodology. In this section, the speaker/teacher is normally an HP engineer who is functioning at the specialist or advisory level of the engineering development pyramid (Figure 7.1). The fourth section is aimed at teaching future trends and developments in the methodology. Speakers for this section come from HP laboratories or universities.

Each of the tutorials is videotaped by a professional crew. The tapes are distributed to the members of the forum as well as the HP R&D engineering community at large. Tapes are viewed by engineers who were unable to attend the expert forum. We find that many people attending the expert forums are more interested in getting together to discuss common interests in a facilitated group environment than attending formal tutorials. Because the tutorials are taped, participants can enjoy this option without sacrificing this aspect of the forum. "Birds of a feather" (BOF) sessions on various related topics allow people to get together to discuss common interests. These sessions run in parallel with the tutorial. Loosely structured and facilitated, BOFs offer an alternative to the person not desiring a formal tutorial on the second day. In addition to the videotapes that are distributed, a white paper is written following each forum. It describes in detail the issues that were addressed. Any engineer at HP may receive a copy of the white paper. Many managers have reported that the white papers have helped them to understand the issues that surround unfamiliar technology.

Expert forums have been a powerful means of technology transfer for HP. Bringing together the right people (engineers on the upper levels of the engineering development pyramid) in this format has generated positive change. For example, a recent expert forum on object-oriented programming methods motivated one project in HP to use Smalltalk on their project. Initially, the project members did not believe Smalltalk could be used in an embedded system. At the forum they learned how Smalltalk could be used in such a system and they applied that knowledge to their project, which finished in half the projected time.

Each of HP's expert forums are evaluated by the members using a detailed questionnaire. Based on the results of the evaluations, the enthusiasm shown by the attending engineers, and the success stories attributable to the expert forums, HP has been extremely pleased with this method of technology transfer.

Consulting

Engineering education at HP has also allowed us to deal with the challenge of moving an engineer to higher levels of the engineering development pyramid (Figure 7.1). Engineers are not only receivers of engineering education, they are also givers. This happens in a variety of ways. In both the leading-edge seminars and the expert forums there is a need for advanced engineers in the company to share their knowledge. We actively seek out the engineers who are perceived to be functioning at specialist or advisory levels. We involve these individuals as speakers or teachers at our seminars and forums.

Specialist/advisory-level engineers are also used in many of the courses for the engineers. While the technology transfer staff develops the curriculum, engineers are often used to implement it by designing the teaching modules. The engineers who teach are generally videotaped so that they do not need to leave their home project each time a course is taught. In addition to courses, seminars, and forums, specialist/advisory-level engineers are used as consultants.

Consulting takes a number of forms. Hewlett-Packard, like most large companies, hires outside consultants in a variety of areas. However, consultants are also found within HP. One way this is done is through a network of engineers. For instance, an HP engineer in Palo Alto may have knowledge and mastery of a technology that is important to engineers in another HP manufacturing site. The engineer/consultant visits the site and gives a day of consultancy to the project team. The receiving site pays for the engineer/consultant's travel expenses and salary for the day.

Hewlett-Packard produces a line of printers that use thermal-inkjet technology to form an image on paper. In the process of transferring this technology to the sites that produce the printers, it was discovered that many engineers and technicians were not familiar with the problems created by electrochemical reactions. An expert in this field was working at an HP Palo Alto site. It was arranged for him to visit the sites needing his consultation, provide an overview of the subject, and give direct consultation to the receiving sites on their specific problems. A number of benefits are illustrated by this example: the engineer/consultant enjoys a higher level of job satisfaction, the consultation was economical for the receiving site, and valuable contacts were established.

Another form of consultation is being piloted in HP. At the beginning of a project a manager may request work-group training. For instance if a project team has determined that they want to use a particular technology, training is provided to the entire project team at the same time. This "just-in-time" training is followed by a consultant joining the project team at the beginning stages of the project to help the team apply the concepts they have learned. This form of consultation was successfully applied to a software team that wanted to do the specification phase of their project using a formal specification language. The entire project was trained in formal specification methods and in the Vienna Development Method (VDM). A consultant then worked with the project team to get them started using the different method.

University Partnerships

Working in partnership with universities has proved to be an effective method of implementing the technology transfer model. HP has formal relationships with such universities as Stanford, Massachusetts Institute of Technology, University of California at Berkeley, and California State University at Chico. Each of these universities provides advanced degrees to our engineers through videotape or satellite instruction. MIT has worked with HP to develop courses that have been taught throughout the company.

Another valued alliance has been with National Technological University (NTU). NTU provides advanced degrees in a number of technical areas through satellite instruction. HP has contributed to the design of NTU's curriculum. In addition, HP has coproduced leading-edge seminars with NTU. This relationship has allowed HP engineers, as well as other NTU subscribers, to hear technical talks that might have been too costly to produce internally.

Conclusion

As HP looks to the future, we would like to see more partnerships with universities and our industrial counterparts. Stronger relationships with universities are already starting to develop. Hewlett-Packard Laboratories has established "science centers" for the purpose of doing joint research with university research departments. The first of these was established in 1988 with Stanford University. Its purpose is to fund research that will be beneficial to both academia and industry.

We would like to see a plan whereby university professors and top HP engineers would exchange places for a year. An experienced HP engineer working for a year in a university engineering department could provide valuable input to curriculum development. A professor working on a project would not only make a strong technical contribution but would gain a deeper understanding of the types of problems industry faces.

We see a need to strengthen our ties to other companies. While this is a sensitive issue, we are seeing movement toward cooperation among competitors. HP's membership in the Microelectronics and Computer Technology Corporation, the Open Software Foundation, and others attests to this.

As a diverse company with a strong corporate culture and value system, Hewlett-Packard presents many challenges to effective technology transfer. The challenge is being met through a multifaceted model that relies on internal HP engineers as well as external resources.

8

Creating and Managing Consortia Cultures through Transitional Episodes

Michael S. Rubin

R&D consortia represent at once a new form of hybrid organization and a new type of interorganizational strategy that is of fundamental importance for the future of U.S. competitiveness. The emergence of these new arrangements is part of the sweeping shift in interorganizational relationships over the past decade, characterized by a remarkable array of multiparty ventures, interfirm alliances, and cross-sectoral partnerships. The exogenous forces contributing to this institutional transformation are multiple, extending from the simultaneous globalization and segmentation of markets to the convergence of multiple technologies in artifacts ranging from the 7J7 experimental aircraft to the "wired family home." We are entering an economic epoch vastly different from the industrial era of the past 100 years—a transformation that will change the structure of markets, institutional hierarchies, and the rules of competition.

The R&D consortium stands out as one of the most significant institutional innovations in this new era, as it provides a bridge between basic research (where the United States continues to be the world leader) and the commercialization of new products (where the U.S. position is slipping across whole industry segments). However, R&D consortia confront substantial managerial, financial, and organizational challenges that limit their effectiveness. These issues will not be resolved without first recognizing that consortia require a distinct technology and a unique ethos to be successful as institutional arrangements.

The "technology," which is perhaps better understood as a "technological gap," we refer to as collaborative strategy. It is at best an underdeveloped technology, being tried and tested in piecemeal fashion within a range of R&D consortia and other interorganizational alliances. The "ethos" that guides these hybrid arrangements is based on multivalence and temporality rather than on the unitary values and rules of continuity and perpetuation that characterize statutory institutions.

Ethos, by which I mean the set of guiding values and beliefs that give form and meaning to relationships between the members of an institution, and technology, by which I mean the way in which artifacts are conceived, developed, and produced through those relationships, taken together represent the foundation of an organizational culture. Therefore to say that in the case of consortia and other multifirm arrangements that both values and methods of production run contrary to what we know about the management and culture of unitary organizations is no small matter.

This chapter was written in response to a major concern regarding the effectiveness of consortia—specifically the ways in which the process of technology transfer is impeded by the cultural differences between the recipient organizations and the producer organization (i.e., the consortium). It is at this critical juncture that the consortium as a temporal creature often terminates—its members generally rejoining their "parent firms" or moving on to other settings. The technology transfer problem extends beyond the "not invented here" syndrome and involves a deeper problem of organizational culture and technology adaptation.

Two interlinked cultural problems need to be solved in order to make consortia effective as organizations. First, the technology of collaboration needs to be clearly defined in contradistinction to both competitive strategies and cooperative arrangements. Second, the episodic and ultimately terminal nature of consortia needs to be fully appreciated relative to the perpetuating bodies that characterize public and private institutions.

The Consortium as a Problem of Organizational Culture

The application of the "culture concept," as borrowed from the interpretive and pragmatic schools of cultural anthropology to explain differences in the orientations and performance of corporations, began in earnest about ten years ago. The "discovery" of organizational culture was a significant event as it signaled a revolution in the way organizations and their managements are perceived, and it has led to changes in the way in which strategies and structures have been shaped.

Four principles dominate the study of corporate cultures. First, it is proposed that corporate cultures have either an implicit or codified set of values and beliefs which legitimate their mission and guide their policies. Second, it is suggested that connecting these values depends upon leaders who champion the organizational mission through their words and deeds. Third, a set of shared stories are developed that exemplify how ordinary members of the organization can become "heroes" through unusual acts that advance the goals of the organization. Finally, there are unique ways of doing things ranging from R&D to customer services which distinguish the organization from its competitors. This "know-how" may be protected as proprietary in the form of intellectual property, market knowledge, manufacturing processes, and the like. Organizational members develop a sense of shared purpose and association through a combination of indoctrination, ongoing "rituals," subscription to the values of the organization, and the development of a role or place within the "daily" routines of the organization.

Within this context the R&D consortium proves to be something of a special problem. First, the consortium consists of the variety of organizational cultures comprising its membership which may from the outset be compatible or in conflict. In this sense, the consortium begins as a hybrid, where members with different values and diverse forms of technological know-how develop a set of guiding objectives characterized by multivalence rather than shared values. This mix of values facilitates multilateral "buy-in" by the consortium's membership. In those instances where a consortium culture does develop univalent principles and purposes it appears to be either the result of acculturation by a dominant organization or enculturation through a consuming commitment to a particular project.

Second, the unusual boundary conditions of the hybrid organization requires an equally unusual form of leadership. As in the case of the unitary organization, the consortium depends on a leader who can inspire its members with a sense of vision and possibility. However, because the consortium is related to a variety of "parents," "investors," or "stakeholders," the leader must also be a skillful politician, able to garner support without making undue compromises and build external coalitions without losing control of the consortium to those influences. In addition, the leader must be skilled in addressing the inevitable internal conflicts that arise between members of the consortium at various points in the R&D process, by balancing the rigor of project management with the vigor of entrepreneurship. In all likelihood, meeting these multiple requirements will involve a leadership team in which the CEO draws upon the liaison, mediator, entrepreneur, and/or project manager to balance his/her own talents. This in turn may lead to some deep ambiguities regarding leadership and power sharing, as for example when a powerful liaison responsible for external relations gains undue influence over the consortium's project agenda.

Third, in addition to the extraordinary boundary management requirements of the consortium, engendering a sense of shared purpose among its members can prove far more perplexing than in the statutory organization. For example, consortium members will inevitably bring different objectives, resources, levels of commitment, and technological skills to the entity creating various asymmetries which can confound their relationships. Another problem is that the essentially episodic nature of the R&D process will almost certainly impact the status, interest, and interrelationships of member organizations at each transition point, causing an ongoing re-sorting of membership roles and positions.

Fourth, the "shared ways of knowing and doing" that characterize the daily culture of corporations may often be illusive for consortium members when various proprietary realms are to be protected from competitor access. Thus, technologies may be modularized, intellectual properties walled off, and special market or production knowledge, which might otherwise influence the product design, may be contained.

Finally, the temporary and ultimately terminal structure of the consortium makes it inherently different from those statutory organizations where the purpose of the corporate culture is to perpetuate values over time. The consortium is more akin to what the anthropologist Victor Turner (1969) referred to as "counter-structure," in describing the nature of ritual processes in which organization members are pulled out of their normal roles and positions to participate in a liminal event, only later to be reintegrated into "normal society" once the rite is concluded. We use this hyperbole to discuss the disconnection between the receiving cultures and the consortium during the technology transfer stage.

Resolving the cultural dilemma of the consortium requires the introduction of an alternative technology or "way of production" which recognizes the simultaneous cooperative and competitive tensions that characterize boundary conditions, rules of membership, proprietary realms, and the shifting nature of relationships through the R&D process. In addition, an ethos that permits a level of multivalence (or what C. M. Turnbayne (1970) called "plurisignificance") is required to incorporate the multiple interests and inherent asymmetries between members.

The first step in this process is to recognize that the consortium is a metaorganization (Figure 8.1.) "Meta" here refers to both the idea of transformation or change (as derived from the Latin) and the concept of a higher order or transcending organization (as derived from the Greek). That is, the consortium is an organization created to transcend the individual capabilities of its member organizations through a transformative process that is ultimately intended to enhance their relative positions. The culture of the metaorganization is based on a transformative technology and a transcendent ethos, that is, a mode of production that serves to transform the modes and orientations of member firms, and an ethos that transcends the particular goals and values of the membership.

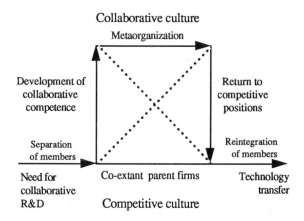

Figure 8.1. The metaorganization as counterstructural transformation.

In the most basic terms, member organizations join the consortium with an ethos either related to the market (i.e., "fulfilling latent needs") or to technological objectives (i.e., "improving the quality of life"). The consortium, by contrast, requires an ethos that co-joins market pull and technological push. To be successful, the consortium must balance the need to create an environment conducive to midterm and longer-range R&D, with an orientation that is responsive to changing configurations and opportunities in the marketplace.

Even more important, member organizations will come to the consortium with either a competitive or cooperative mode of operation. Neither of these modalities will be successful in managing the work of the consortium. A technology of collaboration, which balances the ongoing competitive tensions among members, is essential for successfully traversing the episodic shifts of the R&D process. It is toward this "missing technology" that we turn our attention.

Collaborative Strategy as a Technological Gap

Competitive strategy, which is premised on various assessments of the internal strengths and weaknesses of a firm relative to its competitors and in relation to exogenous market and environmental forces, has over the past fifteen years been developed into a broad array of techniques and orientations. By contrast, as Richard Nielsen (1988) reports, there is no equivalent technological development for interorganizational strategy. One partial exception here involves instances where two or more entities formally merge operations or create a new entity they jointly own. Such cooperational entities are guided more by legalistic instruments during the formative stages than by strategic techniques, but are nevertheless subject to a structured process of creation and governance. Joint operating agreements between area newspapers, joint venture entities, agricultural cooperatives, and merged operations represent such cooperated entities.

In attempting to define the rudiments of collaborative strategy, we decided to begin by distinguishing this emergent technology from its more developed competitive counterpart. In taking this course, five important distinctions can be drawn.

1. Collaborative strategy focuses on the structuring and management of two distinct interorganizational interfaces, where competitive strategy focuses on two distinct levels of the firm.

Competitive strategy focuses either on the business unit level of the firm, where the principal concerns involve the competitive position of product lines or services, or on the corporate level, where issues of product diversification, portfolio management, and overall industry position are emphasized.

By contrast, collaborative strategy focuses on two types of interfaces or boundary management strategies. The first interface involves the interorganizational relationship between the consortium and its parents, investors, or stakeholders. The structuring of specific strategic relationships is essential, as it is the constituent firms that provide access to specific market knowledge, production capabilities, and human capital and financial resources required by the consortium, while it is the consortium that provides access to technologies, know-how, and market opportunities for the individual constituents.

The major obstacle to structuring collaborative strategies at the interorganizational interface is temporal inconsistency. This term, borrowed from finance theory, points out the paradoxes that occur when a firm's goals and objectives are temporally inconsistent with its immediate interests. Absent a collaborative strategy, the temporal orientation of the consortium is likely to depart from that of its parents, exacerbating problems of financial support, intermediate demands, and eventual technology transfer.

The second interface involves the strategic relationship between the participants within the consortium. Here the balance between sharing technologies and know-how and protecting various intellectual properties can cause vexing problems in realizing the enhanced capacities of the metaorganization. As Thomas Roehl and Fredrick Truitt (1987) have pointed out in the case of U.S., Japanese and French ventures in commercial aircraft, a strategy between participating entities must address the "dark side" of collaboration by anticipating the types and quantities of technologies to be transferred, the future competitive position of the participants, the competitive liability and risks caused by early membership withdrawal, the need to wall off certain proprietary technologies, and the need to identify any secondary alliances or relationships of member firms that may jeopardize the position of the consortium.

Collaborative strategy focuses on addressing these two interfaces by establishing a package of joint objectives, identifying interorganizational linkages regarding market sensitivity and R&D paths, setting conditions for parents or investors that reduce temporal inconsistency, and setting conditions for participants regarding intelligence sharing and proprietary protections.

2. Collaborative strategy concerns a negotiated, dynamic process of intelligence bartering, whereas competitive strategy involves a targeted, continuous process of intelligence gathering.

The tension between cooperation and competition between consortium members and the interrelationships between the constituent firms and the consortium itself revolves around the technologies and special forms of knowledge that are to be shared or exchanged. The enhanced capacity of the consortium depends on the success of this bartering process. "Cooperative research" is far too tame a term to capture the inherent difficulties and transitions that characterize the charged negotiations between co-competitors. Participants within the consortium may in one instance be in conflict with one another over the cost or access to a member's technological property, and at another moment jointly be in conflict with the board over restrictions on cooperation they find unreasonable.

Competitive strategy, by contrast, assumes that a firm advances its position by gathering intelligence about, rather than from, its competitors. The development of new capacities within the competitive framework is generally viewed as a decision involving an acquisition, merger, or co-owned joint venture.

3. Collaborative strategy is characterized by a multivocality of goals which attract the support of multilateral constituents, where competitive strategy relies on a more focused, univalent set of goals.

One of the myths concerning the effective management of consortia is that they should be driven by a clarity of vision and a singular set of goals. In the successful consortia we have studied, the goals always appear to be multiple and sometimes are even partially inconsistent. Similarly, the driving mission appears to describe a "region of values" versus a unitary objective.

For example, the Japanese consortium, VLSI, is often pointed to as an example of an interorganizational venture premised on a singular objective—the capture of the merchant DRAM market—and an equally clear set of goals—the creation of a 1,000 DRAM and 1,000 gate logic device. In fact, VLSI organized its activities around a broad array of technological goals, including micromanufacturing, text evaluation, crystal technology, device technology, process technology, and design technology. Thus, this multiparty venture involved a wide spectrum of manufacturing processes and new technological artifacts that held different values for the various member firms within the consortium. In fact, recognizing that this was the case led VLSI to distinguish two types of project participation. "Add Vector Projects" were viewed as ventures in which all parties contributed about equal levels of technological capacity and would benefit about equally. By contrast, "Principal Vector Projects" drew on the capacities of particular firms and therefore conferred unequal levels of benefit.

Indeed, MCC similarly found that pursuit of its longer-term research objectives needed to be balanced by a portfolio of projects that encouraged participation by a broad range of constituent firms with various capacities and interests. Mixing goals to productively integrate the capacities of consortium members represents a special problem in formulating and managing a collaborative strategy.

4. Collaborative strategy is based on a form of contracting which is referred to as relational, in distinction from the neoclassical concept of transactional contracts which apply to competitive strategy.

Competitive strategy is premised on macroeconomic assumptions that firms engage in patterns of transaction in which relationships between suppliers, producers, and buyers is limited to the competitive terms of an exchange. Horizontally related organizations are even more severely limited in their relations by this view due to the threats of anticompetitive practices.

Collaborative strategy, by contrast, depends on what Ian McNeil (1974) defined as a relational form of contracting. Here the participating parties agree to the conditions of a relationship rather than to the terms of an immediate or future exchange. This distinction is a significant one for two reasons. First, the value of the outcome of the R&D effort can rarely be estimated with any kind of precision at the start of the venture. Clearly, certain proprietary technologies can be assessed a value for the purposes of cost pricing a later-stage transfer from one party to others. However, the value of this technology in relation to the value of the resultant technological artifact can only be approximated within some mutually acceptable range.

Second, recognizing the broader conditions and implications of a "relationship" ought to lead to an appreciation that the use of contractual terms based solely on principles of exchange can seriously impede the quality of association between members, reducing the latitude required by the dynamics of an "unfolding" R&D path. Collaborative strategy should therefore incorporate a relational form of contracting to govern changes in the respective interests and positions of participating parties through the R&D process.

5. Collaborative strategy, especially when applied to research and development, must be based on an episodic process, whereas competitive strategy is premised on a continuous process of feedback and adjustment.

Competitive strategy is based on the identification of a set of goals designed to advance the position of a firm, pursued through a sequence of actions, the intermediate results of which are used to adjust or modify the original strategy. This approach assumes that the intentions of the firm will remain fairly constant and that capacities and resources directed to the strategic endeavor are clearly specified. Collaborative strategy, on the other hand, must proceed with the recognition that through the R&D process the intentions, technological capacities, and resource commitments of members may change, sometimes drastically. As E.B. Roberts and A. L. Frohman (1978) have clearly demonstrated, following on the work of Bo B. Klein and W. Meckling (1958), R&D does not involve an incremental or linear sequence of activities, but a series of episodic transition. The stages or phases of the R&D process can be roughly anticipated, but the implications of the transition from one stage to the next cannot.

To address the episodic, discontinuous nature of the R&D process, collaborative strategy must incorporate a number of operational scenarios for each stage of the process, in which the intentions, intermediate goals, and resource commitments of the member firms are likely to change. Where competitive strategies are organized to respond to four stages of industrial growth, collaborative strategies involve a "miniaturization of response" in which alternative modes of collaboration need to be utilized during the four stages of R&D—basic research, prototype development, commercialization, and technology transfer. Larry Hirschhorn of the Wharton Center for Applied Research, who has studied both in-house R&D and R&D consortia over the past decade, has said that unlike the participants in a competitive game who may feel they are engaging in a plan of action, R&D teams often feel that they are engaged in a multiact play, whose narrative is being written in the course of the drama.

Collaborative Culture and Episodic Transitions

The culture of the consortium as a metaorganization is in part a hybrid of the various cultures of its constituent members. However, more significantly it is a culture conditioned by a set of norms and rules that run counter to those of the constituent or parent firms. In addition, as a metaorganization the consortium is a temporary cultural phenomenon, initiated by the actions of certain members who ultimately are to be reintegrated with their parent firms.

The contrary assumptions of the consortium are of course those that relate to collaboration between otherwise competitive parties. The creation of the metaorganization begins with the formation of a core group whose members take on roles that are distinct from those they play in their respective firms. Through a number of discrete stages or episodes consortium members continue to separate from the cultural norms of their respective parent organizations, developing a collaborative set of principles to guide their interactions and relationships. Depending on the degree to which this process of collaborative competence building is successful, a distinctive but temporary culture may emerge. The temporary nature of the consortium is largely consistent with the duration of the R&D process. At each successive episodic transition from early stage R&D to technology transfer, the consortium progressively moves from a collaborative modality to a competitive one. See Figure 8.2.

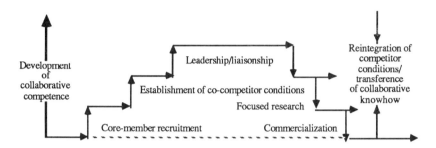

Figure 8.2. Episodic transitions and collaborative competence.

The model of a counterstructural process involving a separation sequence, a liminal or temporal period of activity, and a reintegration sequence is hardly novel. Arnold van Gennep (1960) first suggested this tripartite structure to account for the otherwise paradoxical status shifts and cultural norms that occur during rites of passage. The structural anthropologist Edmund Leach extended this model to account for ceremonies in the modern world, as did Victor Turner (1967) in deciphering why cultures create counterstructural processes. Others have used the same basic structure to explain the "synthetic organizations" that often emerge following a disaster which later disintegrate as social norms and institutions are restored.

What is unique about the counterstructural process in the case of the R&D consortium is the potential for developing new, collaborative competencies which can be partially transmitted back to the participants' parent organizations during the process of reintegration (e.g., technology transfer and utilization). Below are described seven episodic stages that highlight the transitions through which the metaorganization may develop as a collaborative culture, and subsequently reintegrate with competitive cultures of the parent firms. First, the development of the metaorganization as a collaborative structure can be described across three stages of development.

Period of Entrepreneurship and Core-Member Recruitment

The formation of the consortium occurs in relation to a specific technological idea or market opportunity. Typically an entrepreneur, recognizing the need to pool the resources or technological skills of a variety of firms, works to recruit a core group of member organizations around a specific "strategic premise." The strategic premise must have appeal to the various interests and objectives of the core group and provide a rationale for collaboration between the parties.

It is important to note that the strategic premise does not provide the rigor or detail of a strategic plan. To the contrary, the premise allows considerable latitude in interpretation regarding the purposes, means, and structure of the consortium. Through this latitude, core members can further shape a strategy in terms of their respective roles, capacities, and evaluations of the latent opportunities for collaboration.

Expansion and Formalization of Membership

The core group's first task in developing a strategy of collaborative advantage is to identify additional member firms that bring the technological skills, market and production knowledge, managerial talent, and financial strength to support the emerging R&D agenda. Expansion of constituent firms requires the more formal articulation of the criteria and conditions of membership in the consortium be developed. However, in a survey conducted with thirty R&D consortia, the author found that the conditions for membership were often undefined and rarely met the rigor applied to joint venture relationships.

It is at this early formative stage in consortium development that specific conditions and criteria related to technological competencies and properties, market and production knowledge, types and terms of capital commitments, and risk sharing over specific durations or phases can be most productively established as preconditions for collaboration. It is also at this stage that negotiations between consortium participants on competitor conditions ought to take place. These conditions will in part relate to certain initial agreements concerning proprietary boundaries, modularization of technologies, pricing schemes for internal technology transfers, and risk-sharing requirements. In addition, disclosures and agreements on the secondary relationships member firms can maintain with other technology partners outside the consortium need to be incorporated into the conditions that will apply to the collaborative effort.

During this stage the collaborative fit of the member firms and the strategic value of the consortium will become far more concrete. While the difficulty of estimating technological and market benefits will remain illusive, member firms can calculate the various costs of consortium membership against other opportunities for R&D investments and the potential costs of not joining the R&D consortium. It is therefore recommended that during this stage an "exiting process" is provided for firms that find a lack of collaborative fit or strategic value within the consortium.

Leadership, Liaison, and Linkage

In the third stage the consortium shifts its focus from developing collaborative advantage through joint competencies of its members to developing the collaborative competence required to organize and manage a multilateral, multicultural, and multivalent organization. It is at this stage that the leadership of the consortium is selected; relationships with parent firms and other stakeholders become finalized; and the collaborative strategy of the consortium is linked to the organizational structure.

Leadership represents a special problem in the consortium because of the range of skills required to manage a collaborative entity. On the one hand, a consortium often requires a strong chief executive who can inspire collaborative efforts with a bold and energizing vision. On the other hand, the intricate details of the R&D process may call for a certain level of technological expertise; the relationships between constituents may require mediation skills; and the ongoing negotiation with parent firms may call for a savvy organizational politician. In part the actual circumstances of the consortium will shape the criteria for leadership, but it is important to recognize that most consortia will require a leadership team to provide the range of skills required for direction setting, negotiations, and external relations.

A particularly important role is that of the liaison or intermediary who works at the interface between the consortium and the external organizations which contribute personnel, technologies, and/or resources to it. The liaison function actually extends well beyond "managing relationships," as it is essential that the consortium maintain a connection with the marketing, production, and strategic intelligence systems of the parents. Without this connection the commercial potential of R&D products can be seriously jeopardized. It is important to recognize that as a metaorganization configured to conduct transformative R&D, the consortium normally does not have marketing and strategic evaluation systems of its own and can therefore lose touch with external trends and forces. The liaison function is therefore critical in bridging this knowledge gap. It should be pointed out that the precompetitive conditions that apply to horizontal linkages in R&D implicitly preclude in-house intelligence or information sharing which can lead to collusion or parallelism between firms. The liaison function, therefore, needs to be designed with considerable care.

It is also at this third stage of collaborative competence building that the strategy of the consortium will become defined and an organizational structure will subsequently be designed to advance this strategy. Figures 8.3 and 8.4 outline a range of generic collaborative strategies and their relationships to various types of collaborative structures and processes. Unfortunately, there is as yet no substantial case study material on types of collaborative strategy, or more significantly on the relationships between various interorganizational structures and strategic orientations. While further confirmatory work is required, our early research has led us to identify six generic strategies each, of which has implications for the organizational structure and culture of the consortium.

Pooling strategies principally involve the assemblage of either resource capital or intellectual capital to create the foundation for a broad research agenda in a particular technology area. Ironically, pooling strategies are typically faced with a foreshortening of research objectives in response to the demands of investors.

Acceleration strategies are advanced to address an outside competitive threat (presumably international) involving the speed at which an intermediate-stage technology can be transformed into commercial applications.

Sharing or exchange strategies generally involve a small number of parties who engage in the multiparty venture with the explicit intention of trading various types of technological competence, properties, or market intelligence in order to yield mutual benefits vis-à-vis other competitors or competitive alliances. These associations involve considerable risk for the parties in that at the same time they are adding to their own potential competitive advantage they are adding to technological strengths of future competitors.

Stages	Membership Issues
Membership recruitment	Definitions/conditions
Competitor conditions	Boundaries/gates
Leadership	Structure/process
Parallel research	Competence/risk
Focused research	Core/periphery
Commercialization	Prizes/losses
Termination	Disengage/re-engage

Figure 8.3. Transitional stages.

Blocking strategies involve either the joint pursuit of competitive markets ("blocking up") or the purposeful containment of market penetration by competitors ("blocking out"). As in the case of technology sharing/exchange strategies, collaboration exerts a curious tension between the garnering of mutual market advantages by constituents and the creation of more savvy co-competitors.

Linking refers to vertical coupling between producers, suppliers, and buyers, or through consortia structures that are best described as value-added partnerships. These relationships involve a blend of cooperation and competition. On the one hand, as Russell Johnston and Paul Lawrence (1988) have shown, the advantages of vertical linkages between firms from the standpoint of advanced technology and improvements in quality can be quite impressive. On the other hand, any captive relationship, especially a monopsonistic one, can retard the competitive advantages of suppliers and ultimately producers.

Multi-pathing strategies are typically pursued when a technology area involves a mix of hybrid technologies and/or considerable risk in regard to potential discovery paths. Since simultaneous research paths can rarely be mounted by one firm, a consortium structure and collaborative strategy has considerable appeal in such situations. An additional advantage from a macrocompetitive perspective is that multipathing strategies avoid redundancy among a field of competitors. While economists argue that redundancy is desirable because it leads to optional product choices, domestic competitors can be thoroughly overwhelmed by the speed and cost-competitive efficiencies of an international competitive alliance.

Through each episode in the R&D process, the collaborative structure of the metaorganization progressively shifts to a more competitive form of relationship. This transition can jeopardize the collaborative competence developed through the consortium effort if such shifts are not anticipated and incorporated into the collaborative strategy. Using the Turner (1967) model of the counterstructural process, the period of collaborative competence building is equivalent to the "period of separation" in which the rules and norms of the competitive firm are modified to incorporate an alternative mode of interaction. The early stage R&D effort is similarly equivalent to the "liminal period" in which great latitude in the roles and relationships of co-competitors is exercised in contradistinction to the ongoing competitive rules and practices that apply to interactions between the parent firms. Finally, the period of commercial development and technology transfer can be viewed as a gradual reintegration process in which consortium members and intellectual properties are redistributed back to the parents, albeit with added value in the case of successful consortium efforts. In the case of full reintegration, the metaorganizational structure disintegrates, although interorganizational relationships between various parties to the consortium may continue to have an effect on the strategies and actions of firms.

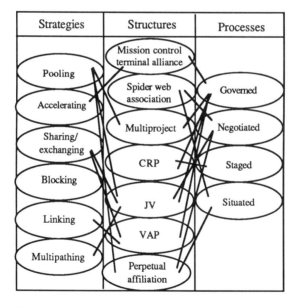

Figure 8.4. Collaborative strategy, structure, and process.

We have found that using this model of the counterstructural process in relation to a four stage R&D process (Roberts & Frohman, 1988) is helpful in portraying the transitory nature of the R&D consortium, and the role of collaborative strategy through each of these episodic transitions.

Early Stage Research

The early stage research effort signals the period during which the collaborative advantages and competencies of the consortium are first tested. Members experience the simultaneous tensions of being representatives of their respective competitive firms and participants in a cooperative R&D engagement. For most consortium participants, structural roles and reporting relationships with their parent organizations are suspended, creating considerable latitude to redefine individual roles and responsibilities within the consortium and to negotiate reporting relationships and communications protocols with the parent.

In our survey of consortia, some respondents described this period as being a time of feeling "betwixt and between," neither part of one's own organization nor fully part of the nascent collaborative effort. Indeed, the fact that during early stage research it is not uncommon to run parallel experiments or projects reinforces the tension between competitive and cooperative behaviors. This is after all a "horse race" in which the value, status, and position of individuals, alternative projects, optional technologies, and constituent firms will be decided.

From the standpoint of collaborative strategy, it is at this stage that the issue of technology transfer should be considered in some detail. First, because this is the point where roles, relationships, and responsibilities of consortium participants will be defined and tested. Second, because it is at this juncture that the consortium can lose touch with the goals, time frames, market intelligence, and production know-how of the parent firms. In other words, following the period of "separation" in which a collaborative modality is developed that distinguishes the strategic orientation of the consortium from that of the constituent firms, it is important to subsequently link the intentions of the "metaorganization" with the expectations and interests that exist among the member firms. Because the consortium as a metaorganization will operate with different values, orientations, and time frames than the parents, a failure to address these differences at this point can obstruct efforts to link with parent firms in the future in regard to technology transfer, licensing agreements, technology utilization, and marketing rights. This is a cross-cultural problem, in which the nascent collaborative culture of the consortium can get out of touch with the competitive cultures of the constituent firms.

Transitional, Focused, or Prototype Research

A successful early-stage research venture leads to the identification of a preferred technology route or development path. For the consortium, this transition will inevitably result in a shift in the relative influence and position of its members, including the possible exiting of some members. In other words, a partner with a technological competence that was viewed as central to the consortium at the initiation of the venture may find that its technology is of secondary importance during the prototype development stage and that its position within the consortium has slipped. Alternately, a major financial contributor may find that the results of the early-stage research do not meet its original expectations and may look to reduce or withdraw future support.

The resulting asymmetries and changes in relationships between consortium members has a direct impact on collaborative strategy. It is at this stage that projects may need to be restructured and/or repartitioned to address the concerns of the participants in the face of a new opportunity/risk environment. Similarly, from the standpoint of the parent firms and other stakeholders, it is likely that any alteration of current positions and future expectations will result in efforts to renegotiate terms and conditions.

It is important to recognize that by this time the consortium may have been in existence for two to four years, and if it is successful it will have developed a strong organizational culture which members may identify with more strongly than they do with their parent firm. Indeed, in our survey of consortia close to 80 percent of the respondents reported feeling a higher level of commitment to the consortium. The perceptions within the consortium about the consequences of this stage of research may therefore vary significantly from the perspective of firms operating at the consortium's interorganizational level. Within the consortium there may be strong support across the membership for a particular project regardless of its differential value for various constituents, while at the interorganizational level the competitive orientations of the firms may dominate over commitments to the consortium.

The liaison function is of great importance in providing a cross-cultural link between these two organizational realms. It is in the role of the liaison to act as an interpreter, translating the concerns and perceptions of the consortium regarding the technological opportunities to the parent firms and the risk tolerances and expectations of various parents to the consortium leadership. It is likely that through this second stage of the R&D process a new core and periphery structure will develop within the consortium as the result of changes in member roles, positions, and relationships, and due to changes in the interests and expectations of parent firms. This structural change may have direct implications on the leadership structure of the consortium. From the standpoint of collaborative strategy, this is a point at which leadership transitions should be anticipated.

Commercialization

The next stage in a successful R&D consortium centers on the precommercial testing or commercial development of a prototype technology. Current restrictions on horizontal forms of cooperative research place limitations on the degree to which allied parties can pursue commercial development jointly. Without discussing this important boundary condition, it is at the onset of this stage that consortium members begin to resume their competitive orientations. This point marks the "reintegration period" in the counterstructural model at which time individual members of the consortium begin to rejoin the parent firm, and access to interorganizational technologies and resources becomes more limited.

Technology transfer can of course take place at this stage in regard to rights to development, licensing arrangements, component use, and joint-access agreements. What is significant from the perspective of collaborative strategy is that technology transfer occurs at a point in which the consortium begins to terminate as a metaorganization with a distinct identity and cultural presence. The process of technology transfer, which depends on the producer organization (the consortium) communicating the utility, value, and application of a new technology to the recipient organization, can therefore be jeopardized unless a clearly defined phasing of the "reintegration process" is incorporated within the collaborative strategy.

Consortium Termination and Technology Transfer

The final episodic transition of the consortium involves its termination and the subsequent partitioning of its technological products and assets. Termination is rarely as definitive as the term suggests. In multiproject consortia, termination may refer to the conclusion and disassemblage of a single project. In a contract research agency it may involve terminating one multiparty venture while continuing work on another. Nevertheless, to one extent or another the multiparty venture loses its status as a separate entity, and the venue created for collaboration no longer exists. In horizontal consortia, termination may involve negotiating certain ongoing linkages that cannot be dissolved. For example, if facilities are jointly owned by consortium members a residual linkage will persist between the parties. It is important in such cases to fully recognize that new conditions must be applied to these post-termination relationships, as the parties of interest are now competitors.

Another interesting post-termination situation occurs when the consortium members perpetuate their relationship after the original R&D agenda has been completed. As Roehl and Truitt (1987) reported in the case of the Boeing-JADC consortium, the difficulties of forging and managing a multiparty relationship may lead parties to "stick to the devil" they know, rather than pursue new alliances with the unforeseen difficulties they may bring. From the standpoint of collaborative strategy, it is important to appreciate that at termination the understandings that conditioned the relationship are partially or wholly separated. It therefore is incumbent upon the parties to reassess and refigure these conditions rather than perpetuate a phantom relationship. Returning one last time to Turner's (1967) counterstructural model, it is important to note that during the period of reintegration the participants do not simply return to their former roles, but are accorded a new status based upon the special knowledge or insight gained during the liminal period. By analogy, one would hope that in the reintegration process the parent recognizes the added value participants bring back to the firm from this experience and rewards these risk takers for their efforts. Unfortunately, this rarely appears to be the case and represents one of the more curious problems of "technology transfer." That is, people who have gained both new technical competencies and collaborative skills are not generally viewed as part of the transfer of technological value and often find they have lost status and possibly even a place in their parent organization.

Conclusion

Collaborative strategy is an underdeveloped technology which needs to be further defined and developed as an accessible technique to guide multiparty R&D ventures. However, it is only through the actual experiences of R&D consortia that these principles can be tested and made concrete. An essential element in the process of technology transfer is the transference of the collaborative skills and practices developed by the consortium to the parent firms in the form of retained competencies. Only in such a case will the consortium truly perform the function of a metaorganization in transforming the capabilities of the constituent firms.

References

Anderson, Eric. "Assessing the Performance of a Joint Venture," RHJ Working Paper WP 88–04.

Astley, W. Graham. "Toward an Appreciation of Collective Strategy," *Academy of Management Review,* March, 1984, pp. 526–35.

_____ and C. F. Fombrum. "Collective Strategy: The Social Ecology of Organizational Environments," *Academy of Management Review*, August 1983, pp. 576–86.

Axelrod, Robert. *The Evolution of Cooperation.* New York: Basic Books, 1984.

Baxter, William F. "Antitrust Law and Technological Innovation," *Issues on Science and Technology*, Winter 1985.

BNA Antitrust and Regulation Report. "U.S. Department of Justice Antitrust Guidelines for International Operations," vol. 54, no. 1369, June 9, 1989.

Bowman, Edward H. "Strategy Changes," Working Paper WP 88–02.

_____ and Declan Murphy. "Government Regulation, R&D, and the Pharmaceutical Industry," Working Paper WP 84–09.

_____ and Dileep Hurry. "Strategic Options," Working Paper WP 87–02.

_____ and Joseph T. Mahoney. "Types of Competition: A Framework for the U.S. Chemical Industry," Working Paper WP 87–08.

Brahm, Richard A. and W. Graham Astley. "Constrained Exploitation: A Reevaluation of Governance and Performance in Joint Ventures," Working Paper WP 88–01.

Bresser, Rudi K. and J. E. Harl. "Collective Strategy: Vice or Virtue," *Academy of Management Review*, November 1986, pp. 408–27.

Camerer, C. "Redirecting Research in Business Policy and Strategy," *Strategic Management Journal*, June 1986, pp. 1–15.

Clemons, Eric K. "MAC—Philadelphia National Bank's Strategic Venture in Shared ATM Networks," Working Paper WP 88–13.

_____. "When to Lead, When to Follow, and When to Go Your Own Way: An Analysis of the Available Options when Investing in Information Technology," Working Paper WP 87–19.

Conner, Kathleen Reavis. "Paradigms and Strategy," Working Paper WP 88–25.

Contractor, F. J. and Peter Lorange. "Competition vs. Cooperation: A Benefit/Cost Framework for Choosing Between Fully-Owned Investments and Cooperative Relationships," *Management International Review*, Special Issue 1988, p. 5.

Deal, Terrence E. and Allan A. Kennedy. *Corporate Cultures: The Rites and Rituals of Corporate Life*. New York: Addison-Wesley, 1982.

Dimanescu, Dan. *The New Alliance: America's R&D Consortia*. Cambridge, Mass: Ballinger Publishing Company, 1986.

Eccles, Robert G. "The Quasifirm in the Construction Industry," *Journal of Economic Behavior and Organization*, vol. 2, 1981, p. 335–37.

Economist. "IBM-Rolm," September 29, 1984, pp. 75–76.

Economist. "Competing by Collaborating," June 21, 1986, pp. 19–20.

Evan, William M. and Paul Olk. "R&D Consortia: A New U.S Organizational Form," Working Paper WP 89–03.

Feder, Barnaby J. "Turning On the Research Switch: The Utilities' Research Arm Shows the Pluses and Pitfalls of Cooperation," *The New York Times*, May 14, 1989, 6F.

Felker, Lansing. "Cooperative Industrial R&D: Funding the Innovation Gap," *Bell Atlantic Quarterly*, Winter 1984, vol. I, no. 2.

_____ and Susan L. Miller. "The Need for Cooperative R&D: Case Studies," Industrial Technology Partnerships Program, U.S. Department of Commerce, June 1985.

Freudenheim, Milt. "Drug Makers Try Biotech Partners," *The New York Times*, September 30, 1988, D1.

Fusefield, Herbert I. and Carmela S. Haklisch. "Cooperative R&D for Competitors," *Harvard Business Review*, November–December 1985, pp. 60–76.

Galbraith, C. and D. Schendel. "An Empirical Analysis of Strategy Types," *Strategic Management Journal*, vol. 4, 1983, p. 153–74.

Gennep, Arnold van. *The Rites of Passage*, London: Routledge and Kegan Paul, 1960.

Gerlach, Michael. "Business Alliances and the Strategy of the Japanese Firm," *California Management Review*, Fall 1987.

Gobeli, David H. and William Rudelius. "Managing Innovation: Lessons from the Cardiac-Pacing Industry," *Sloan Management Review*, Summer 1985, p. 29.

Gomes-Casseres, Benjamin. "Joint Venture Instability: Is it a Problem?" *Columbia Journal of World Business*, Summer 1987, p. 87.

Gray, Barbara. "Conditions Facilitating Interorganizational Collaboration," *Human Relations*, vol. 38, no. 10, 1985, p. 911.

Gullander, Staffan. "Joint Ventures and Corporate Strategy," *Columbia Journal of World Business*, Spring 1976, p. 104.

Hamel, Gary, Yves L. Doz, and C.K. Prahalad. "Collaborate with Your Competitors and Win," *Harvard Business Review*, January–February 1989, pp. 133–39.

Harianto, Fred and Johannes M. Pennings. "Technological Innovation through Interfirm Linkages," Working Paper WP 88–24.

Harrigan, Katherine. "Strategic Alliances and Partner Asymmetries," *Management International Review*, Special Issue, 1988, p. 53.

Helm, Leslie and Alison L. Cowan. "IBM Wins the Key to Japan's High-tech Labs," *Business Week*, August 19, 1985, p. 48.

Huallachain, Breandan. "Cooperative R&D Ventures: An Analysis," Northwestern CUED Institute, prepared for Economic Development Administration, U.S. Department of Commerce, October 1986.

Jacques, Laurent L. "The Changing Personality of U.S.–Japanese Joint Ventures: A Value-Added Mapping Paradigm," Working Paper WP 86–02.

Johnston, Russell and Paul R. Lawrence. "Beyond Vertical Integration—The Rise of the Value-Adding Partnership," *Harvard Business Review*, July-August 1988, pp. 94–101.

Killing, Peter. "Technology Acquisition: License Agreement or Joint Venture," *Columbia Journal of World Business*, Fall 1980, p. 38.

Kilmann, Ralph, et al. *Gaining Control of the Corporate Culture*, San Francisco, Calif.: Jossey-Bass, 1985.

Klein, B. and W. Meckling. "Applications of Operations Research to Development Decisions," *Operations Research*, 1958, vol. 6, p. 352–63.

Kogut, Bruce. "Country Competitiveness: International Patterns in Innovation and Imitation over Time," Working Paper WP 87–16.

_____ and Harbir Singh. "Entering the United States by Acquisition or Joint Venture: Country Patterns and Cultural Characteristics," Working Paper WP 85–12.

_____ and Nalin Kulatilaka. "Multinational Flexibility and the Theory of Foreign Direct Investment," Working Paper WP 88–10

_____. "International Strategy as a Game of Boules: Changes in Comparative and Competitive Value-Added Chains in the World Economy," Working Paper WP 84–14.

_____. "Joint Ventures and the Option to Acquire," Working Paper WP 88–19.

_____. "Joint Ventures: A Review and Preliminary Investigation," Working Paper WP 86–07.

_____. "Cooperative and Competitive Influences on Joint Venture Stability under Competing Risks of Acquisition and Dissolution," Working Paper WP 86–09.

Kraar, Louis. "Your Rivals Can Be Your Allies," *Fortune*, March 27, 1989, pp. 66–76.

Levine, J.B. "How IBM is Getting the Most Out of Rolm," *Business Week*, November 18, 1985, p. 110.

Lorange, Peter and Gilbert J. B. Probst. "Joint Ventures as Self Organizing Systems: A Key to Successful Joint Venture Design and Implementation," *Columbia Journal of World Business*, Summer 1987, p. 71.

Lyles, M.A. "Learning among Joint Venture Sophisticate Firms," *Management International Review*, Special Issue 1988, p. 85.

MacMillan, Eric, Murray B. Low, and Edward H. Bowman. "Event Study Methodology and Invisible Assets," Working Paper WP 88–22.

McNeil, Ian. "The Many Futures of Contract," *Southern California Law Review*, 1974, 47, 5:691–816.

Marcom, J. "More Companies Make Alliances to Expand into Related Businesses," *The Wall Street Journal*, November 8, 1985, p. 1.

Martin, Edwin M. and Carol H. I. Martin. *Financing Research and Development*, Chicago: Commerce Clearing House, 1987.

The New Climate for Joint Research, Conference Proceedings, May 13, 1983, U.S. Department of Commerce, Office of the Assistant Secretary for Productivity, Technology, and Innovation, US GPO, 1984.

Nielsen, Richard P. "Cooperative Strategy," *Strategic Management Journal*, vol. 9, 1988, pp. 475–92.

———. "Industrial Policy: The Case for National Strategies for World Markets," *Long Range Planning*, October 1984, pp. 50–59.

———. "Cooperative Strategy," *Strategic Management Journal*, vol. 9, 1988, p. 475–92.

Norris, William C. "Cooperative R&D: A Regional Strategy," *Issues on Science and Technology*, Winter 1985.

Ohmae, Kenichi. "The Global Logic of Strategic Alliances," *Harvard Business Review*, March–April 1989, pp.143–54.

Osborn, Richard N. and C. Christopher Baughn. "New Patterns in the Formation of U.S./Japanese Cooperative Ventures: The Role of Technology," *Columbia Journal of World Business*, Summer 1987, p. 57.

Ouichi, William G. and Michele Kremen Bolton. "The Logic of Joint Research and Development," *California Management Review*, vol. 30, no. 3, Spring 1988.

Peet, James W. "Technology Alliances: An Interview with James Grant," *The McKinsey Quarterly*, Autumn 1988, p. 58.

Pennings, Johannes M. "Strategic Inertia and Turnaround: An Action-Oriented Framework of Strategic Change," Working Paper WP 87–04.

Perlmutter, Howard V. and David A. Heenan. "Cooperate to Compete Globally," *Harvard Business Review*, March–April 1986.

Pucik, Vladimir. "Strategic Alliances, Organizational Learning, and Competitive Advantage: The HRM Agenda," *Human Resource Management*, Spring 1988, vol. 27, pp. 77–93.

Roberts, E.B. and A.L. Frohman. "Strategies for Improving Research Utilization," *Technology Review*, March/April 1978.

Roehl, Thomas and Fredrick Truitt. "Stormy Open Marriages are Better: Evidence from U.S., Japanese, and French Cooperative Ventures in Commercial Aircraft," *Columbia Journal of World Business*, Summer 1987, p. 87.

Rubin, Michael S. "Sagas, Ventures, Quests, and Parlays: A Typology of Strategies in the Public Sector," *Strategic Planning*, 1989, pp. 84–105.

Shan, Weijan and Kanoknart Visudtibhan. "Cooperation as a Competitive Strategy in Commercialization an Emerging Technology," Working Paper WP 88–08.

Smith, Lee. "Can Consortiums Beat Japan?" *Fortune*, June 5, 1989, pp. 245–54.

Thorelli, H. B. "Networks: between Markets and Hierarchies," *Strategic Management Journal*, vol. 7, 1986, pp. 37–51.

Turnbayne, C. M. *The Myth of Metaphor*, Columbia: University of South Carolina, 1970.

Turner, Victor. *The Forest of Symbols*, Ithaca, N.Y.: Cornell University Press, 1967.

_____. *The Ritual Process,* Ithaca, N.Y.: Cornell University Press, 1969.

Walker, Gordon. "Network Analysis for Cooperative Interfirm Relationships," Working Paper WP 87–14.

_____. "Strategic Sourcing, Vertical Integration, and Transaction Costs," Working Paper WP 87–18.

Welter, Therese L. "Hurling R&D Gaps," *Industry Week,* vol. 234, no. 1, July 13, 1987.

White, Lawrence C. "Clearing the Path to Cooperative Research," *Technology Review,* July 1985.

Wolff, Michael F. "Teams Speed Commercialization of R&D Projects," *Research Technology Management,* September–October 1988, p. 8.

Part III

Policies and Procedures for Technology Transfer

9

Transferring Technology in R&D Consortia: Effective Forms of Interorganizational Relations

William M. Evan
Paul Olk

Technology transfer is a seemingly clear and straightforward idea but upon closer inspection it is baffling. The concept was first introduced by development economists in their search for ways to accelerate the development of Third World countries. By transferring Western technology they were convinced that they could help less-developed countries (LDCs) make a transition to the twentieth century. This, alas, did not happen. An array of cultural and educational obstacles, not to mention political corruption, frustrated their worthy objective. Multinational corporations, in the course of establishing subsidiaries in LDCs, frequently found themselves defending their operations against various criticisms by claiming that they were transferring Western technology and training host-country employees in the use of modern technology. In this respect, managers of multinational corporations have been appreciably more successful than development economists engaged in large-scale projects financed by the World Bank and the International Monetary Fund.

On the face of it, technology transfer, in the context of an interorganizational relationship such as an R&D consortia, should have a much better chance of success than developmental economics or multinational corporations, both of which operate in a complex macrosocietal context. In practice, however, the process is fraught with difficulty. The development and exploitation of new technology are becoming increasingly important for the success of companies and countries. Because of the trend of shorter technology life cycles, companies are less able to maintain a competitive advantage over rivals for an extended period of time. As a result, a premium has been placed not only on developing new technologies but also on developing products from these technological findings. To aid in this process, R&D consortia have been formed; for an R&D consortium, two critical functions are the research and development of technology and the transfer of the findings to member companies. This chapter explores the issue of how to transfer technology to the member company. Problems arise in a number of areas which hinder the transfer. To transfer technology successfully requires active participation by both consortium researchers and company researchers. Unless a concerted effort is made on both sides, the transfer will be problematic. This chapter examines some of the problems of technology transfer and the mechanisms managers find most effective for transferring technology.

Background

Research and development consortia have been legal since U.S. antitrust laws were changed in 1984. With the passage of the National Cooperative Research Act (NCRA), the U.S. Congress opened up an avenue for improving the innovation ability of U.S. companies. The law permitted the formation of R&D consortia. In passing the law, proponents asserted that a consortium structure was necessary to maintain U.S. competitiveness. The impetus for this new organizational form was a concern that with declining corporate R&D expenditures, U.S. companies would soon lag behind foreign companies in technological innovation (Wright, 1986). It was argued that rapid changes in the complexity of technology necessitated an increased investment in research by U.S. companies for them to remain competitive. However, the costs—in some cases over a billion dollars—the risks, and the complexity associated with this type of research made it likely that only the largest of companies could make such an investment. And, as Jorde and Teece (1988) argue, these large companies are not well suited for developing the needed "radical" innovations. Thus, a call was made for the new governance structure. Consortia are an alternative to internal venturing, informal technology transfer, and licensing arrangements.

Companies are increasingly drawn to cooperative R&D as they become more dependent on external science and technology for business maintenance and growth (Haklisch, Fusfeld, and Levenson, 1986). The principal reasons for joining a consortium are to achieve economies of scale, to share the risks involved in an innovation, to set a standard for a new technology, to share complementary knowledge, and to help protect proprietary rights and recoup investments from developing technology with "leaky property" rights (Ouchi and Bolton, 1988). While there are many advantages to membership, an R&D consortium has potential disadvantages stemming from increased uncertainty associated with this organizational form (Pisano, 1988; Von Hippel, 1988; Evan and Olk, 1990).

Regardless of the motives for joining a consortium, common to all is the expectation that the findings of the research will be transferred to member companies. While technology is also being transferred from the company to the consortium and may also be transferred from the consortium to nonmembers, the important link is from the consortium to the company. If this link does not exist, or is weak, the consortium is likely to fail. As a result, we focus our attention on this direction of transfer.

Policies and Barriers to Transferring Technology

Transferring technology in consortia is viewed by both managers of companies and of consortia as being more difficult than in traditional organizations. From survey questionnaire data, which we will describe in detail below, both consortia and company managers reported that technology transfer was more difficult than in traditional organizations. To understand why it may be more difficult, we will explore some of the barriers to technology transfer and then some of the policies that have been developed in order to overcome these barriers.

Barriers

One way to view technology transfer is as a form of interorganizational relationship. As has been noted in the literature, organizations transmit information between and among themselves through a variety of mechanisms. Besides commonly known ones such as published reports, personal contacts, and trade associations, other mechanisms used for relaying information include personnel transfer (Baty, Evan, and Rothermal, 1978), interorganizational data systems (Stern and Craig, 1978), board of directors meetings (Pfeffer, 1972), and contractual or joint venture arrangements. Regardless of the mechanism, the overriding concern for most companies is how to restrict technology transfer as much as possible. The loss of technology to a competitor may result in a competitive disadvantage. Hence, policies and regulations such as the development of patents, restrictions on employment practices—such as preventing a researcher from working on a similar topic for a competitor—and antitrust provisions have developed.

In a consortium arrangement, however, technology transfer involves the opposite problem: how to encourage or facilitate the transfer of information between two organizations—the consortium and a member company. This problem is difficult because of a number of functions inherent in a consortium arrangement. Specifically, we will discuss the influence of NIH or the "not invented here" syndrome, the timing of the transfer, and the complexity of the body of knowledge involved.

The NIH phenomenon is well known in research settings. While not unique to consortia, it is still troublesome. Smilor and Gibson (1991) report that consortium employees believe NIH at the member companies to be a significant barrier to technology transfer. Consortia managers (in interviews we have conducted) describe a situation in which a sense of competition may develop between consortium and company researchers. As a result, when a finding from a consortium is transferred to the company, it may be ignored by member company researchers.

The second problem, timing of the transfer, is related to NIH. Some consortia managers we have interviewed found that transferring information to a company without the company being prepared for it resulted in the company not using the information. The explanation is that the company was not conducting research identical to the consortium research. When the findings were presented, company researchers were not able to incorporate these findings into their research projects.

The third type of barrier to technology transfer is a function of the complexity of the body of knowledge and technology. Akin to research at universities and research labs, some research findings conducted at consortia are not easily transferred through reports or telephone calls. Such information is often difficult to transmit through conventional mechanisms. Only by conducting the research can an individual understand the technology. This places a premium on the behavior and interaction of personnel conducting the research: to ensure the transfer of technology to member companies, researchers themselves must, in effect, become liaison personnel or "information carriers" in the process of boundary spanning (Aldrich and Herker, 1977).

Policies

Many consortia have developed policies that attempt to overcome barriers to technology transfer. In general, these policies try to create a tighter coupling between the companies and the consortium through a continuous flow of information. Mechanisms used in this transfer can be divided into two types: personnel-related mechanisms and technology-related mechanisms.

Personnel-related mechanisms. Liaison employees are one of the mechanisms most often cited as being important for technology transfer. Liaisons are employees of the company who work at the consortium full-time but are responsible for transferring back information to the company. Characteristics that have been cited as important for effective liaison functioning are that the liaison personnel know what is important for member company researchers, as well as that they be respected in the company and that the information transferred goes to the appropriate individuals.

A related policy for transferring technology is to use company "assignees" to the consortium. These are employees of the companies who temporarily work at the consortium. Unlike liaisons, they are not as responsible for keeping the company aware of the consortium's progress as they are for conducting the research. Their job is to develop an understanding of the technology being developed. The tenure for assignees is typically between one and two years, although shorter assignments are not uncommon. During their stay at the consortium, assignees help the consortium conduct the research. At they end of their assignment, they are expected to return to the member company with the knowledge they acquired and continue the research there.

Additional personnel-related policies of transferring technology include board meetings and technology advisory board (TAB) meetings. The former usually consists of top-level company managers who meet to discuss overall consortium goals and progress. The latter meet to discuss more day-to-day operations of the research, set technical guidelines and make decisions about how to conduct the research. Technology advisory boards tend to have more knowledge of the consortium's technology and may represent a better technology transfer mechanism than board meetings. A final personnel mechanism used by some consortia is what are known as company visitation days. This consists of a company sending several researchers and executives to the consortium for a day or two to visit with consortium researchers and managers.

Technology-related mechanisms. Other mechanisms besides personnel-related ones are used to transfer technology. Many consortia have policies concerning the use of videotapes, research reports, and conference calls to transfer information to companies. However, while these mechanisms are effective for transferring simple information, they are less successful for transferring complex technology.

In addition, many consortia have policies for licensing any technology developed. If member companies do not want to produce the technology, the consortium may offer it to nonmembers. Profits from the licensing arrangement will be funnelled back to the consortium.

Transferring Technology

Following our discussion of some of the policies for transferring technology which we gleaned from interviews with consortia managers, we now turn to information collected via two survey questionnaires addressed to consortia managers and company managers.

R&D Consortia Population

The consortia population for this survey consists of the consortia listed with the Justice Department under the NCRA of 1984. In exchange for exemption from some of the antitrust regulations, consortia file with the Justice Department the stated goal of the consortium, and the company members, as well as any subsequent changes in either of these. Since the passage of the NCRA of 1984, as of August 1989, 137 cooperative ventures have been formed. Whether this number reflects the actual number of consortia is open to interpretation. For example, two of the earlier consortia, Bell Communications Research, Inc. (BELLCORE), and Petroleum Environmental Research Forum (PERF) have subsequently filed additional research projects (Bellcore–27; PERF–5) which include nonmember companies. While these are listed as separate consortia, managers of the consortia do not believe these are separate strategic efforts. Alternatively, other consortia, such as MCC, have several ongoing projects with different members but are listed as only one consortium. Also, two research organizations, Motor Vehicle Manufacturer's Association (MVMA) and Southwest Research Institute (SWRI) which are not listed as consortia, oversee numerous research projects but have only registered a subset of the projects (MVMA–15; SWRI–6) under the act.

After taking into account these considerations the consortia population was reduced to 84 organizations, to which we sent survey questionnaires.[1] To the approximately 1,100 companies we identified as members of these 84 consortia, we sent a second questionnaire asking many of the same questions as in the consortium questionnaire. We did not send this questionnaire to universities or to domestic or foreign governments, and we did not send questionnaires to some companies for which we were unable to locate addresses, companies that had merged with other companies or that were no longer operating, and companies that had a policy against responding to questionnaires. As a result, a total of 684 company questionnaires were mailed out.

The response rate for the consortium survey was 52 percent; 44 questionnaires were returned, of which 42 were usable. For the company questionnaire, we received 207 responses or a response rate of 30 percent. Taken together, these questionnaires provide data on 68 of the 84 consortia (78 percent). For 28 of the consortia we have both company and consortium responses, for 12 consortia we have only consortia responses, and for 28 of the consortia we have only company responses.

Technology Transfer Mechanisms

Included in each of the questionnaires were items asking respondents what mechanisms were most effective for transferring technology to the company from the consortium. Table 9.1 contains a list of the mechanisms used and the percentage of consortia that use each type.[2] For the 68 consortia, most use a variety of mechanisms to transfer technology. Most consortia use research reports, technical demonstrations, conference phone calls, employee visits to the consortium, board meetings and visits to companies for transferring technology. Slightly fewer consortia use video tapes and employee transfers to transfer information.

Table 9.1. Percentage of Consortia Using
These Mechanisms to Transfer Technology
(N = 66)

Technical Reports	96
Video Tapes	74
Telephone Conference Calls	88
Technical Demonstrations	87
Board of Directors/Technical Advisory Board Meetings	87
Consortium Employee Visits to Company	87
Member Employee Visits	87
Member Employee Transfers	80

For each of these mechanisms, we also asked respondents to rate technology transfer effectiveness of 1 to 5 on a scale, 1 = not at all effective and 5 = very effective. For this analysis, we separated the responses by organizational affiliation—consortium vs. member company. The results are presented in Figure 9.1.

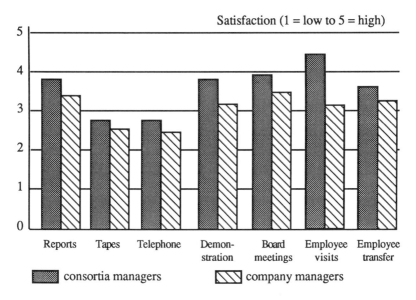

Figure 9.1. Mean satisfaction with technology transfer mechanisms.

For the consortium managers, company employee visits to consortia are clearly perceived to be the most effective mechanisms for transferring technology. Still considered effective, but less so, are research reports, technical demonstrations, employee transfer, and board meetings. Finally, mechanisms found to be least effective are video tapes and conference phone calls. These findings suggest that as the mechanism becomes less personal, the satisfaction with the mechanism decreases. One anomaly to this trend is that company employee transfers are not viewed as effective as are consortia visits to companies. While we do not know for sure the reason for this difference, one interpretation is that the expense of transferring employees (i.e., losing a researcher for an extended period of time) may offset the advantages the mechanism has for transferring complex information.

For company managers, there is less of a difference in their satisfaction among the mechanisms. Research reports, technical demonstrations, technical advisory boards, researcher visits, and liaisons are all found to be effective mechanisms for transferring technology. Less effective are video tapes and conference phone calls. While the pattern found for consortia managers is present, the company managers are generally less satisfied with technology transfer mechanisms. For example, the mean scores of consortia managers' satisfaction for employee visits was 4.31, while for company managers the scores were 3.20.

Finally, it is important to note that the correlation between company and consortia scores was not significant. So, while there is general agreement as to whether or not these mechanisms are used, there does not appear to be a congruence among companies and consortia on their satisfaction. From this analysis we conclude that what consortia managers believe are effective mechanisms for transferring technology is not identical as what company managers believe.

Technology Transfer Satisfaction and Mechanisms

In an attempt to understand the importance of both managers of the company and managers of the consortium working on technology transfer, we turn to data on satisfaction with overall technology transfer of the consortium. As with the mechanisms for transferring technology, we asked both consortia managers and company managers how satisfied they were with the transfer of technology to companies (on a scale of 1 to 5 where 1 = not at all satisfied and 5 = very satisfied). We then collapsed the responses into two groups: low satisfaction (for responses 1, 2, or 3) and high satisfaction (for responses 4 or 5). The result of this grouping is that 17 consortia had a high satisfaction score and 20 had a low satisfaction score (five were missing this data). For company managers, 94 had a low satisfaction score and 64 had high satisfaction score (31 were missing this data). Because we are interested in the situation in which both sides are satisfied, we combined these two scores and created a transfer satisfaction topology. Figure 9.2 shows the four types.

Company Satisfaction

		Low	High
Consortium Satisfaction	Low	Neither satisfied (31)	Company satisfied (22)
	High	Consortium satisfied (46)	Both satisfied (26)

Frequency of each type in parentheses

Figure 9.2. Typology of satisfaction with technology transfer.

The types are low company–low consortium satisfaction (N = 31); high company–low consortium satisfaction (N = 22); low company–high consortium satisfaction (N = 46); and high company–high consortium satisfaction (N = 26). Because of missing data in one variable or another, the sample size is now 125.

We now turn to an analysis of the relationship between satisfaction with technology transfer and the mechanisms used to transfer. Table 9.2 shows the relationship.

Table 9.2. Relationship of Transfer Mechanisms to Satisfaction Typology
(1 to 5 scale, 1 = not very satisfied, 5 = very satisfied)

Transfer mechanism	Neither satisfied	Consortium satisfied	Company satisfied	Both satisfied
Technical reports*	3.53	4.11	3.45	3.75
Video tapes	3.04	2.76	2.90	2.88
Telephone conference calls*	2.13	2.75	2.70	2.87
Technical demonstrations	3.60	3.84	3.49	3.73
Board of directors/technical board meetings*	3.36	3.60	3.22	3.62
Member employee visits*	3.48	3.73	4.00	4.09
Member employee transfers*	3.29	3.73	3.22	3.96

*Analysis of variance test finds a significant variation among the four types for the transfer mechanism ($p \leq .10$).

As the table indicates, there is a significant difference among the types in their satisfaction with technical reports, telephone conferencing calls, board meetings, member employee visits, and member employee transfers. When both consortium and company managers are satisfied with overall technology transfer, there is greater satisfaction with personnel-related mechanisms. When both are dissatisfied there is greater satisfaction with technical demonstration. When only one of the managers is satisfied with technology transfer, the satisfaction with personnel-related and technology-related mechanisms is mixed. In general, these results confirm the importance of personnel-related mechanisms for transferring technology.

In interpreting these results, we need to clarify an assumption we are making about satisfaction with technology transfer. Earlier we argued that technology transfer requires active management by both company and consortium managers. While a measure of satisfaction does not directly reflect the effort made by managers, we assume that satisfaction is in part a function of effort. That is, managers who work at transferring technology will be more satisfied than those who do not. Furthermore, we assume that when both consortia and company managers are satisfied, an effort is being made by both sides of the transfer.

Evidence supporting this assumption is found in relating overall technical performance of the consortium, an indication of the usefulness of the technology transferred, to satisfaction with technology transfer. The results, presented in Table 9.3, find that having both company and consortium managers satisfied with technology transfer is associated with better overall performance of their consortium—a subjective evaluation of consortium research along several technical dimensions, excluding technology transfer.

Table 9.3. Relationship between Technology Transfer Satisfaction Type and Performance

Consortium performance*	Satisfaction Type				
	Neither satisfied	Company satisfied	Consortium satisfied	Both satisfied	Total
Poor	10	3	15	1	29
Good	21	19	31	25	96
Total	31	22	46	26	125

Chi-Square Statistic 10.30 (significant at $p < .02$, $df = 3$).
*Performance is derived by taking the composite score of company and consortium managers' satisfaction with several technological elements.

Additionally, we should note that we controlled the transfer-satisfaction relationship for the effects of four factors that may affect technology transfer: age of the consortium, focus of research (e.g., basic, applied), consortium form (e.g., for profit, nonprofit), and location of research (e.g., new organization, university). The results of this analysis found that age was not a factor in satisfaction and neither was organization structure nor location of research. However, there was a variation in satisfaction with the focus of the research. Managers involved in consortia addressing prototype and standard development research were much more satisfied with the technology transfer: company and consortia managers involved in basic research were significantly less satisfied with technology transfer than were other managers. Managers of other types of research operations (e.g., applied research, pilot plant development) were not significantly different in their satisfaction.

Discussion

As our analysis suggests, technology transfer in consortia is a complicated process and requires active management on both sides of the process. Our study also shows that some of the mechanisms are viewed as more effective for transferring technology than others. Specifically, face-to-face mechanisms such as visitations and board meetings are viewed as more effective than impersonal mechanisms such as technical reports, phone calls, and video tapes. Finally, our study finds a relationship between perceived consortium performance and satisfaction with technology transfer.

These results have several implications for technology transfer and consortium management. The first implication is that to manage effectively technology transfer in a consortium requires an effort by both sides of the transfer relationship. When an effort is made by one side but not by the other, technology transfer mechanisms tend to be more passive. Because of the complexity of the technology involved in consortium research, these may not be the most effective mechanisms. While we cannot impute causality from our data, we can speculate that the technology transfer mechanisms chosen will influence the satisfaction with the transfer. Furthermore, we find that choosing the less people-intensive mechanisms leads to less satisfaction.

The second implication is the importance of face-to-face contact for transfer. While this is probably a function of the complexity of the technology, it is probably also a function of the value of human interaction for processing and remembering information. Consequently, it is important that managers build into the technology transfer policies, efforts and opportunities for face-to-face interaction.

Conclusion

Based on this analysis, several conclusions can be reached. First, policies and mechanisms for technology transfer need to be bilateral; that is, both sides of the transfer process have to actively manage the process. Unless there is compatibility between company and consortium policies, barriers of resistance will emerge. Second, the personnel involved in the transfer makes a crucial difference. Because of the importance of personnel-related mechanisms for transferring technology, the so-called boundary spanning positions—liaisons, transferred employees—are significant in successfully transferring information to a member company. The people who occupy these positions are critical for the process. Third, top company and consortium managers should be involved in the process. While they may not actively transfer the technology, they need to ensure that the policies and mechanisms reflect a commitment to technology transfer. Technology transfer is the pay-off for a consortium. Unless there is a whole hearted commitment, the consortium will not succeed.

In conclusion, the adage that was stated by a consortium manager of "you only get out of this (consortium) what you (company managers) put into it" is only half correct. It should be reworded to state, "you only get out of this what *we* put into it," the *we* being the consortium and company managers. Only by actively pursuing technology transfer will consortia attain the success they have been designed to achieve.

Notes

1. Since we could not locate addresses for two consortia we subsequently mailed out 82 questionnaires. When the consortium did not have a central organization, we sent the questionnaire to the highest-ranking officer of the consortium which we could identify at a member company.
2. There was a high concordance between company and consortia respondents for whether or not a mechanism was used to transfer technology. Consequently, where consortia responses exist, they are used. But when only company responses were available, we report these responses.

References

Aldrich, H. and D. Herker. "Boundary Spanning Roles and Organization Structure," *Academy of Management Review*, vol. 2, no. 2, April, 1977, pp. 218, 231.

Baty, G., W. Evan, and T. Rothermal. "Personnel Flows as Interorganizational Relations,". in *Interorganizational Relations,* ed. W. M. Evan, University of Pennsylvania Press, 1978, pp. 202–18.

Evan, W. M. and P. Olk. "R&D Consortia: A New U.S. Organizational Form," *Sloan Management Review,* vol. 31, no. 3, Spring 1990, pp. 37–46.

Haklisch, C., H. Fusfeld, and A. Levenson. *Trends in Collaborative Industrial Research.* New York: New York University Press, 1986

Jorde, T. and D. Teece. *Innovation, Cooperation and Antitrust,* Monograph. University of California–Berkeley, 1988.

Ouchi, W. and M.K. Bolton. The Logic of Joint Research and Development, *California Management Review*, Spring 1988, pp. 9-33.

Pfeffer, J. "Size and Composition of Corporate Boards of Directors," *Administrative Science Quarterly*, vol. 17, 1972, pp. 218–28.

Pisano, G. "The Governance of Collaborative Innovation: Equity Linkages in the Biotechnology Industry." Working Paper. University of California–Berkeley, 1988.

Smilor, R. W. and D. V. Gibson, "Technology Transfer in Multi-Organizational Environments: The Case of R&D Consortia," *IEEE Transactions on Engineering Management*, vol. 38, no. 1, February 1991, pp. 3–13.

Stern, L. and C. Craig. "Interorganizational Data Systems," in *Interorganizational Relations*, ed. W. M. Evan, University of Pennsylvania Press, 1978, pp. 409–22.

Wright, C. "The National Cooperative Research Act of 1984: A New Antitrust Regime for Joint R&D Ventures," *High Technology Law Journal*, 1986, pp. 135–93.

Von Hippel, E. *The Sources of Innovation*, New York: Oxford Press, 1988.

10

A Consortium-Based Model of Technology Transfer

Larry Novak

In 1986 the Science Board of the Department of Defense convened a task force to study the effect of U.S. semiconductor dependency. It issued a report in February 1987 which concluded that the implications were serious for the nation's economy and security. The most significant finding of the task force was that U.S. technology leadership in semiconductor manufacturing was rapidly eroding and that this had serious implications for the nation's economy and immediate and predictable consequences for the Defense Department. The report further concluded that action must be taken to:

- Retain a domestic strategic semiconductor production base
- Maintain a strong base of expertise in the technologies of circuit design, fabrication, materials refinement and preparation, and product equipment.

In response to this call for action, SEMATECH was formed. SEMATECH[1] began in Santa Clara, California in 1986 and moved to Austin, Texas in April 1988. As consortia go, SEMATECH is unique to the American experience. It was formed to focus on advances in semiconductor manufacturing, not basic research. SEMATECH is concerned with continuous transfer of semiconductor manufacturing tools and methods. We do incremental transfers all through the program.

SEMATECH is not the total answer. SEMATECH's annual budget is only $200 million per year, and you cannot plug all the holes or the broken links in the semiconductor industry chain with just $200 million per year. By contrast, the European JESSI organization is spending in excess of $5 billion. One half of SEMATECH's funding is provided by the Defense Advanced Research Projects Agency (DARPA), which is in the U.S. government Department of Defense. This funding is matched by the consortium's member companies. SEMATECH's mission is easy to say, but it is extremely difficult to accomplish. The objective is to provide the U.S. semiconductor industry with the domestic capability for world leadership in manufacturing. To accomplish this SEMATECH must continually transfer to its members the knowledge related to the design, development, and demonstration of semiconductor manufacturing equipment.

SEMATECH is only one resource that is being used to accomplish this mission. SEMATECH will only be effective to the degree that we can get synergism from all entities depicted in Figure 10.1. SEMATECH's technical communication organization, which reports directly to the executive office at SEMATECH, is responsible for all the technology transfer and all technical communications with members, suppliers, universities, and the U.S. government. See Figure 10.2. All the elements necessary to do effective technology transfer are in the technical communications organization. Two key elements in this transfer process are marketing and engineering. Engineering is responsible for working with the engineers as they go through the development and demonstrations of the different programs and gathering the resultant technical information. As a career engineer, I believe one of the common characteristics of engineers is that they do not want to tell anybody what they're doing. They kind of want it to be a surprise and if it is not a surprise, they do not want anybody to know about it. Such lack of communication results in problems for SEMATECH since we have to tell the member companies, because it's their money, not only our successes but also our failures, so they can learn from those failures. Engineering is probably one of the most critical parts of the organization. There are conflicts. But an effective engineering organization is critical to getting technology out of a company like SEMATECH.

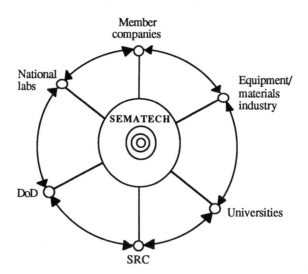

Figure 10.1. SEMATECH as a part of a larger set of entities needed to improve the U.S. semiconductor industry.

Marketing is responsible for getting the information to SEMATECH's members. I mean marketing in the true sense of the word. It's where you have customers and you go out and see what they want and you try to give it to them. You do not know if they're going to take your product unless someone puts money on your side of the table for the products you are providing. Marketing is an essential part of this technology transfer.

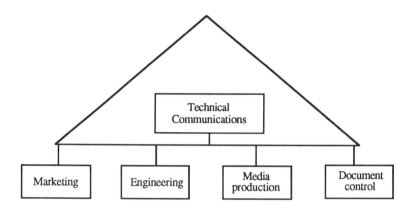

Figure 10.2. SEMATECH's technical communications organization.

SEMATECH grew so quickly that some of its functional areas started and grew parallel with others, even if they were supposed to be reporting to each other. Figure 10.3 shows how SEMATECH is governed. As SEMATECH was formed, a communications structure was established with its member companies to ensure close communications and collective goal setting. Technical advisory boards (TABs) correspond to SEMATECH's technology focus areas. Each of the member companies is represented on each one of the focus TABs. These TABs provide input both to SEMATECH on an advisory basis and to the executive technical advisory board (ETAB). They are also proving to be a key method of technology transfer: not only from SEMATECH to the focus TABs, but also between the members in the focus TAB. The ETAB takes input from the TAB and provides technical advice to SEMATECH's management. It reports directly to the board of directors from which strategic direction for SEMATECH emanates. The transfer mechanisms in Figure 10.3 are intellectually driven, and each has its own specific charter. The technical communication organization is chartered with coordinating these transfer mechanisms. The technical communications organization is a central entity in SEMATECH which is chartered with all of the communications with the outside world. This involves technology transfer.

The real key to effective technology transfer at SEMATECH lies in the ability to get the technology into the hands of those in the member companies who use it to create and produce products. See Figure 10.4. A major part of the transfer equation involves a network of member-company full-time assignees which comprise approximately one-half of the engineers at SEMATECH. The assignees contact is a primary receiver who resides at the parent companies and who is responsible for transferring the technology into the appropriate base of end-users within the company. Also critical to the success of this model are the member-company technology transfer managers, who are responsible for ensuring effective transfer into their respective companies. They also form a hard-working focus technical advisory body (FTAB) within SEMATECH's Technical Communications Organization. FTAB, which was formed in August 1988, has been critical in setting up the methodology for technology transfer at SEMATECH.

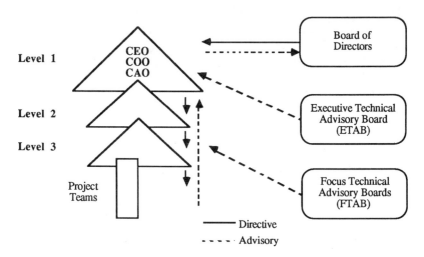

Figure 10.3. The governance of SEMATECH.

Workshops have been a key instrument in getting SEMATECH up and running and in setting the direction of the consortium. Seminars and symposia are used to transfer information to and from SEMATECH members and suppliers. SEMATECH has had two major transfer sessions: one on communications and one on the facilities. The facilities transfer was probably the most successful, because there are member companies using that information as they start putting up new factories. They can look at SEMATECH and if there's something good they can use it. If we've made mistakes, then they can also learn how to avoid those mistakes. At the first transfer session we had, over a three-day period in November 1988, we averaged about ninety company members, and that did not include the follow-up trips.

One of the important things at SEMATECH that we are kind of late in discovering is the importance of the documentation of visits by members. We want to understand why certain member company representatives are visiting SEMATECH. Basically the visits by company members to SEMATECH averaged about 140 per month for all fourteen companies. Around mid-1988 we started visiting the member companies at their sites. We believe that SEMATECH had transferred a lot of information, and the only way we could determine how that information was being used was to visit with marketing personnel at their site. SEMATECH representatives walked into the member company offices and saw rooms full of documents. We knew we were in trouble. The purpose of these visits was to facilitate the transfer of information to the companies, and they have been an eye opener and probably one of the most successful and fruitful things that we have done.

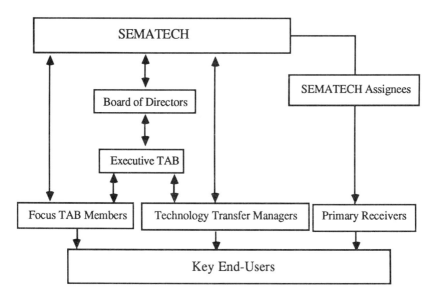

Figure 10.4. Member-company technology transfer from SEMATECH.

User groups, which SEMATECH has experimented with, are direct confrontation meetings where we put the member company that is actually using the particular piece of equipment in a room with the supplier and we go through what is wrong with the supplier's equipment. It is an amazing meeting. About a year and a half ago we probably could not have held such meetings, but they have resulted in some excellent results.

SEMATECH is transferring about twenty documents per month or over 200 to date. We plan to move from paper as a primary media to electronic transfer in the first half of 1990. Still, the most important element of transferring technology is the human element: one-on-one and hands-on experience. See Figure 10.5. Documentation is vital, but it is only a record of the transfer and not the transfer itself. At SEMATECH we involve people in a variety of ways, many of which are in the context of formal meetings. Figure 10.6 shows the types of meetings at SEMATECH and who is involved in each. The emphasis is on the integration of the member companies, supplier companies, universities, and national labs. Over 150 such meetings had been held at SEMATECH by 1989.

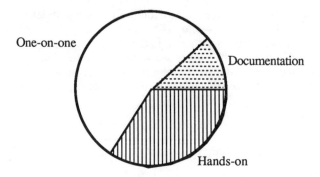

Figure 10.5. Most effective means of technology transfer at SEMATECH.

Following is a list of the elements of successful technology transfer.

- *Industry participation.* It is imperative that SEMATECH membership reflect the mainstream of U.S. industry and that they actively participate in the consortium from the very beginning. SEMATECH's members comprise nearly 80 percent of the U.S. semiconductor production base.

- *Focus activities.* It is important that SEMATECH always focus its efforts on those things best done by consortia as opposed to individual member companies. These are high-risk, large-payback programs. Information exchange is needed to provide the focus of members and suppliers on developing critical equipment and on strengthening key areas.

- *Customer focus.* Perhaps most importantly, SEMATECH must realize it is producing a product. We have customers, and the product is knowledge. If SEMATECH's customers are not involved both early and continuously in the technology transfer process, we may not give them what they need or want. We may also transfer the technology in an unacceptable fashion, and the member companies may not want to receive it. Listen to the customer. This is critical. SEMATECH must use a well-organized means of collecting and using customer feedback in our planning and evaluation process.

- *Serve the customer.* We have to be flexible enough to change when change is required in the fast-paced industry of semiconductors. SEMATECH must learn to be realistic by setting realistic goals. In the semiconductor business, where six months is often a difference between market success and complete failure, the U.S. cannot afford to lose a race with our competitors. SEMATECH must execute crisply and quickly. It cannot be late. Our goal in technical communications at SEMATECH is to operate in Quadrant B, Figure 10.7. The customer wants it and gets it. In Quadrant C the customer does not want it and does not get it. The other two quadrants simply irritate the customer.

	Member Companies	Suppliers	Universities and National Labs
Executive TAB	•		
Focus TABs	•		
Advisory Councils	•		
Workshops	•	•	•
Seminars	•	•	
Symposia	•	•	•
User Groups	•	•	
Transfer Sessions	•		
SCOE Reviews	•	•	•
Presidents' Days		•	

Figure 10.6. The types of meetings at SEMATECH and the participants.

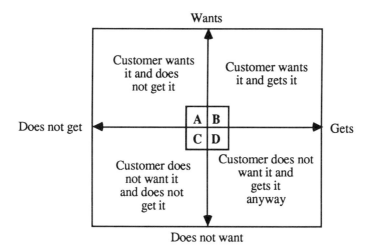

Figure 10.7. SEMATECH's customer window.

Note

1. SEMATECH is an acronym made up of the first letters of the words semiconductor, manufacturing and technology. The fourteen-member companies of SEMATECH are AMD, ATT, Digital Equipment, Harris, Hewlett-Packard, Entel, IBM, LSI Logic, Micron, Motorola, National Semiconductor, NCR, Rockwell, Texas Instruments. DARPA is the fifteenth company.

11

Technology Transfer at the Software Productivity Consortium: A Partnership Approach

Claude DelFosse

In many American homes, the video-taped stories we watch on our VCRs provide an excellent source of entertainment and escape. But to those of us who earn our livings in technology transfer, these leisure-time devices tell a different story—one from which there is no escape. From Sony to Sanyo to Samsung, the VCR is a constant reminder that a company's ability to create technological innovation is not enough to assure product success. Firms like those just mentioned have proved the case in the worldwide economic language we might call "market share." As many companies in the United States have only begun to realize, success in the laboratory is just the beginning of a successfully deployed product.

The Software Productivity Consortium is in the innovation business. More specifically, we are an initiative of some of the leading companies in the U.S. aerospace, defense, and electronics industries, established to accelerate the pace of large-scale software development among our member companies and other customers, and to improve the quality, suitability, and extensibility of their software systems (see Figure 11.1). To do this, the consortium creates, acquires, integrates, and transfers software products and technologies, including processes, methods, tools, and services. Consortium products are currently used in dozens of real-time, mission-critical systems being developed by our members for use in a wide variety of government and military programs.

Conceptual thinking about the organization dates to the late 1970s. In that time frame, leaders in both government and industry identified a threat not only to the nation's ability to create the very high-speed integrated circuits and related hardware technologies necessary for next-generation advances in aerospace systems, but also a challenge to its ability to generate the software needed to exploit such devices. By 1983 a group of aerospace executives began organizational discussions about a software consortium, culminating in a memorandum of understanding in the following year. The consortium was incorporated in 1985. Also in that year, the organization developed both its initial technical and business plans and selected a temporary headquarters location in Reston, Virginia. The consortium's organizational phase ended in June 1986, and it began to assemble a superior technical staff. In December 1988, the company moved to its permanent facilities, a state-of-the-art 60,000-square-foot office building in the Center for Innovative Technology complex, Herndon, Virginia.

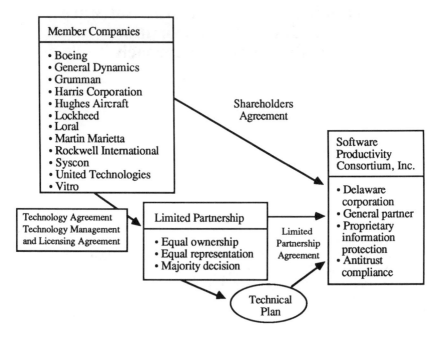

Figure 11.1. A partnership of industry leaders.

The consortium technical program complements the software productivity initiatives of the federal government, the Defense Department and military branches, academic and private institutions, commercial software companies, and the in-house programs of member companies. The consortium technical program has several distinguishing attributes, however: first, it creates fully supported software engineering processes, methods and tools for its customers; second, it targets real projects and transfers products for use in the short term, as well as providing support for more dramatic midterm improvement; third, it focuses longer-term technology investigation in high-payoff areas, with risk higher than that usually acceptable in the internal programs of our member and customer organizations.

In some ways, it would make my job easier to say our technologies, like the VCR, get packaged in a black box and cabled to the back of a television set. They do not. As mentioned above, we deal in the incorporeal world of process, method, and software. Our goal is the transfer of new processes, methods, and tools for developing large-scale and highly complicated software-dependent systems. Our process improvements will be supported by new development methods and tool sets for automation of essentially manual, repetitive, and often error-prone tasks.

All parties to the consortium recognize that new technology remains inert without transfer strategies to move it out of the lab and into the field, and all act as collaborators in the design and development of our technologies. We have developed a very "proactive," multifaceted technology transfer strategy, with process, marketing, and service dimensions. The first, process, integrates and superimposes a technology transfer process over the technology development life cycle. The second creates innovation awareness, understanding, and persuasiveness. The third supports a company's decision to adopt the technology, guides its implementation, and nurtures a continued commitment to product use. At each step, software engineers from our member companies advise and counsel us on our plans and activities, and educate others in their organizations about our program. In short, we view technology transfer as a people-to-people activity, or as I like to call it, a contact sport.

Because the customer set we serve represents literally thousands of engineers and other technical professionals, we attempt to multiplex our efforts through a common set of transfer "agencies" (as Figure 11.2 indicates, these agencies can be located at the corporate or divisional levels of an organization):

- Board member
- Technical Advisory Board (TAB)
- Transfer sites
- Distribution points
- Software committees
- Assignees
- Technical Advisory Groups (TAGs).

A board member is a member company's senior management interface to the consortium. Each company has one representative to the board of directors. The board meets six to twelve times yearly to review program progress and vote on major management and policy issues.

The Technical Advisory Board (TAB) is composed of key technical representatives from each of the consortium's member companies, and plays an instrumental role in advising and directing the consortium's technical program.

The consortium supports two *transfer sites* per member company. A transfer-site organization is a focal point for member company interactions with the consortium. Typical transfer site activities include product and prototype evaluations, document and plan reviews, product distribution and installation, internal adaptation, and internal training and support.

Distribution points are individuals, strategically placed within their respective member companies, who act as transfer agents for consortium products, including guidebooks, technical reports, tool-set evaluations, technical plans, and other types of documentation. Distribution points maintain the mailing lists needed to circulate this material. As a result, they both service requests and identify critical recipients who may otherwise not have knowledge of or access to the consortium products. We serve over 150 distribution points.

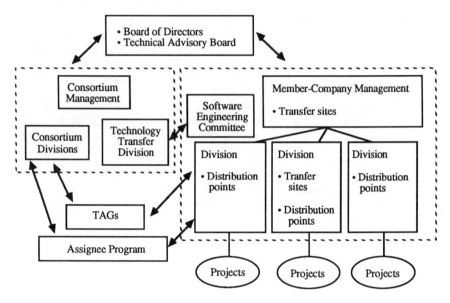

Figure 11.2. A common set of transfer mechanisms supports the technology partnership.

Software committees (Software Engineering Process Groups, Software Engineering Working Groups) are a critical focal point for consortium communication with its member companies. These are groups within the membership which help set internal software policies and procedures, evaluate advanced technologies, and set technical directions. Given this orientation, software committees are a natural interface for the consortium program.

Assignees are software engineers and other computer scientists from the member companies who work at the consortium for periods of up to two years. Assignees interact as full members of the consortium technical staff and, upon their return to the member company, continue to provide consortium program liaison and advocacy.

Technical Advisory Groups (TAGs), composed of member company engineers and computer scientists, work with each of the consortium's major technology projects. A TAG provides industry perspective and advice to consortium project teams, offers valuable input on actual member-company environmental and corporate culture conditions and constraints, and serves as in-house liaison between their respective companies and the consortium's technical program. Because the TAGs offer early involvement in the formative stages of technology, the consortium believes these groups form an important bridge for successful transfer.

In addition to appropriate transfer mechanisms, the consortium has developed a framework for technology transfer which accommodates the process, methods, and tools to be conveyed, the environment in which the transfer takes place, and the steps required for successful adoption.

Our goal is to establish *"key uses"* of our technologies by our members, where the usage of consortium products is considered, by the customer, to be vital to the success of the project. Perhaps unique to the Software Productivity Consortium, this "key use" metric has already helped us identify dozens of instances where our products are so used, and provides an effective measurement of our successes in "satisfying the customer." This notion of "key uses" serves multiple purposes: (1) In a resource-limited environment, it enables us to concentrate our technology transfer efforts where they have the greatest leverage potential. (2) Key uses are selected from major highly visible national programs, usually involving many member companies and many people, and thus increase the number of users and attendant benefits received. (3) Experience has shown that it is better to have a few good uses rather than a multitude of mediocre ones.

The "accommodation" phase of our technology transfer activities recognizes the special characteristics of the technology being transferred. This means benefits of the new technology must be thoroughly identified and communicated. The technology itself must be built with an eye to how it can be fit smoothly into existing processes—as well as how it must evolve to meet out-year requirements. Ease of learning and use are part of the accommodation process. The technology should be introduced in several digestible bites—big bang approaches run the risk of big backfires. Our technology transfer framework helps us address each of these issues.

The framework also helps us to recognize when and how our technology transfer activities must be tailored to specific environments, or to support the varying sizes, characteristics, and behavior of member organizations. Variant factors within the corporate cultures of each member company, which our transfer mechanisms must address, include:

- "Passive" versus "active" membership participation;
- Mistaken perception of the consortium as an outside *supplier*, not as a technology *partner;*
- Extreme variations among member companies in the delegation of management-level responsibility for corporate participation in the consortium program;
- Centralized versus decentralized receptor sites and user organizations;
- Multiple company funding models;
- Project versus overhead receptor status;
- Intracompany competition;
- Intracompany communications problems.

Again, the framework allows us to "stretch" to identify and accommodate particular member company needs. In addition, an active government affairs program furthers our understanding of the future software technology requirements of our customer's customer—often, the federal government. By understanding the needs of federal and defense agencies, and advising them on the advantages offered by our technologies, we create a "pull" demand for our products which our members are well-positioned to meet.

Technology adoption takes place in a series of steps, each of which involves close collaboration with as wide an audience among our membership as possible.

Awareness creates an initial impression and predisposition; *understanding* builds the base of information necessary to evaluate the innovation; *persuasion* is the mapping of product features and benefits to customer requirements; *decision* is the decision to adopt or reject; *implementation* addresses product installation and early use; *confirmation* is the reaffirmation or rebuttal, after initial usage, of the original adoption of a decision; *satisfaction* is the recognition of benefits derived (e.g., the identification of "key uses" of our technologies).

The consortium transfer program is composed of several joint initiatives, conducted in close partnership with our members, designed to satisfy these adoption steps (summarized in Table 11.1). I would like to describe each of these briefly and conclude with some lessons gleaned from almost five years of first-hand technology transfer experience in a consortium environment. The initiatives fall into four principal service categories: general program information and communications, technical transfer, technology support services, and customer feedback. These activities are incorporated into our plans for the development and evolution of our technologies and are continuously reviewed by our customers to ensure that our transfer activities are ongoing and successful.

General program information services keep our member companies abreast of major technical program issues and activities. We attack this communication challenge in several ways.

Member Company Liaison

Every consortium member company is assigned a Technology Transfer Manager (TTM). The TTM is a senior consortium staff member whose role is twofold: first, this individual must understand the fit between consortium products and services and the programs and needs of the member; next, this person works to facilitate the interactions necessary for the successful use of our technology. Each TTM visits member company sites several times a year to discuss various facets of the consortium program and to collect member company feedback. Often this feedback takes the form of actionable items. The TTM assures a prompt and relevant response from the consortium to all member-company requests and suggestions. To a large extent, the TTM is a member company's' advocate within the organization, working to reflect a member's needs and concerns in the overall technical program.

The Software Productivity Consortium *Quarterly*

Published four times a year, the *Quarterly* features the major initiatives of the consortium's technical program, including news and information about guidebooks, products, prototypes, video tapes, and related offerings. The publication covers complicated technical subject matter in a straightforward easy-to-read and easy-to-understand style, making it an ideal mechanism for quickly "coming up to speed" on consortium activities. The *Quarterly* also features success stories of applications of consortium technologies in the customer environment. It is distributed to more than 10,000 subscribers.

Table 11.1. Consortium Transfer Matrix

Initiatives	Awareness	Understanding	Persuasion	Decision	Implementation	Confirmation	Satisfaction
General Information and Communications							
Liaison	•	•	•				
Quarterly	•	•					
Videos	•	•					
Executive Communications	•	•	•				
Technical Transfer							
Guidebook	•	•	•	•	•		
Prototypes		•	•	•	•		
Strategic Alliances				•	•	•	
Workshops/Seminars	•	•					
Clearinghouse	•	•	•	•	•		
Advisory Groups	•			•	•		
Speakers	•	•					
Technology Support Services							
Pilot Project		•	•	•	•	•	
Education and Training	•	•	•	•			
Consulting Support	•	•	•	•	•		
Demonstration Center	•	•					
Customer Feedback							
Surveys						•	•
In-depth Questionnaires						•	•

Note: Understanding through Confirmation fall under *Member-Company Collaboration*.

Video Series

The consortium has successfully used video to communicate, to tens of thousands of software engineers, three classes of information about its program: marketing, tutorial, and documentary. Of recent note are our management summary videos, which provide the viewer with a quick summary of how key consortium products (e.g., ADARTS, Dynamics Assessment Toolset, Systematic Reuse, etc.) can address their immediate and longer-term needs. The current consortium video catalog lists nearly fifty distinct titles. As appropriate, the consortium's in-house distribution lecturer series is recorded, and edited tapes offered to member companies' video distribution points. New titles are listed in each issue of the *Quarterly*. All told, the consortium has shipped more than 3,000 videos to its members.

CEO Executive Newsletter, Technology Transfer Progress Reports, and Benefit Reports

Successful technology transfer results from a mixture of top-down management direction and bottom-up perception of technical benefit. To ensure proper management awareness of the value and benefits of the consortium program, a number of top-level management communications mechanisms have been instituted.

The *CEO Executive Newsletter* addresses top-level topics, such as technology needs arising from government policy or agency directive, or uses of consortium technology in major programs of our members. It is distributed every two or three months to senior-level executives such as company and/or division presidents, senior vice-presidents, etc.

The *Technology Transfer Progress Report* is intended to ensure that every engineering manager within the corporation knows of consortium-related activities throughout his/her company. It is distributed every two months to more than 400 vice-presidents of engineering and software engineering managers.

Benefit Report, prepared once a year, provide a summary to each member company of all the technology transfer activities, including usage of technology, participation in consortium events, attendance at education and training events, etc., for the preceding year. It also quantifies, in dollars, the actual and potential benefits derived from improved productivity and quality resulting from the application of consortium processes, methods, and tools. The two major messages are: without technology transfer and use, there is no benefit; conversely, the greater the extent of use, the greater the benefits.

Technical Transfer Services seek to provide our member company engineers and other project personnel with a context for understanding our processes, methods, and other technologies. This set of services supports the need to transfer technology in "digestible" increments. To this end, we have formulated several service offerings.

Technical Reports, Guidebooks, and Case Studies

Consortium technologies are either composed of, or supported by, extensive documentation, which provides step-by-step instruction on advancing the maturity of our members' software engineering practices. In keeping with the consortium's focus on process improvement many of our "technologies" are, in fact, knowledge about the software engineering process, the correct usage of appropriate method, and the mixed use of commercially available tools as needed. This knowledge, packaged in technical reports, guidebooks, and case studies, constitutes, in many cases, the actual consortium product being transferred. Examples include our

Evolutionary Spiral Process, addressing the complete software life cycle; our ADARTS design method for Ada real-time systems; our *Ada Quality & Style* programming guide; and other products for software measurement, verification, and reuse.

Technical reports represent the results of our initial investigations into new technologies or methods for software engineering, and set forth our planned directions for developing new products. These reports are carefully reviewed and analyzed by a broad spectrum of member-company engineers and project managers; their feedback is incorporated into our preliminary product plans and is gathered throughout the product-development process.

Based on this foundation, we begin to develop products that meet the concerns expressed during review of our technical reports. Chief among these are consortium guidebooks, comprehensive collections of guidelines for using the best available processes, methods, and tools to solve a given problem in software development. Consortium guidebooks are flexibly structured to support the process improvement efforts of organizations at varying states of software maturity, and help software engineers employ advanced techniques and practices on a daily basis. Their value comes from providing a "defined, repeatable" process for the most difficult of software engineering tasks, which otherwise are often left to individual intuition or experience. Consortium guidebooks also provide for ongoing collection and evaluation of information about the software engineering process, and "lessons learned" during particular projects. This allows each customer to build a corporate data base of experience in software development and support, a critical asset if processes are to be improved.

Case studies, detailing the application of the consortium technology being transferred in actual systems development, serve to highlight particularly important aspects of the use of new technologies. Increasingly, consortium guidebooks contain case studies describing our own activities to validate that technology, often performed in concert with other academic and industrial organizations. As a recent example, our Evolutionary Spiral Process guidebook is accompanied by a case study describing the validation of the process on an actual, real-time systems development project conducted with George Mason University's Center for Excellence in C^3I Systems.

Prototypes

As a further collaborative mechanisms with our members, and consistent with our philosophy of providing incremental support for the processes and methods our members need, the consortium offers various levels of prototypes. Typically, to receive a prototype copy, member company recipients come to the consortium and participate in workshops, demonstrations, and familiarization classes and training. This quick education sets the contextual issues straight: what the tool does, which direction it will evolve, and ground rules by which it should be evaluated. To ensure usefulness, prototypes are built using rigorous software development and testing procedures and are released only after reaching demanding quality standards.

Strategic Alliances

Automated support tools further enhance the transfer of methods and processes. Similarly, prototypes do not transfer well to real "live" projects without the support of commercial-strength software tools. In order to provide this support and further extend the usability of our technologies, the consortium is working in close partnership with commercial vendors of computer-aided software engineering (CASE) tools. Through this strategic alliance program, vendors incorporate consortium processes, methods, and prototypes into their already popular tools, and provide the resultant products to the consortium member companies. This creates a "win-win-win" situation. The consortium benefits through speeding up technology transfer by incorporating its technologies into products already familiar to its customers; the member companies benefit by receiving fully supported automation products; and the commercial vendors benefit through the addition of state-of-the-art technologies and an expanded marketplace.

Technology Transfer Clearinghouse

In addition to software products, the consortium distributes technical reports and other documentation related to its program through its clearinghouse organization. The technical report series covers a wide variety of software engineering topics, including process improvement, reuse, Ada, software design, performance assessment, domain analysis, measurement, and methods. For example, the consortium has published tool evaluation reports which scan the computer-aided software engineering marketplace and provide comparative results on a number of commercially available tools. It has also developed a methodology to evaluate CASE tools. Another unique document, *Ada Quality and Style: Guidelines for Professional Programmers,* published by Van Nostrand Reinhold, originated as a consortium technical report and is still offered to member companies through the clearinghouse. (*Ada Quality and Style* is now recommended by the DoD Ada Joint Program Office [AJPO] as "the suggested Ada style guide for DoD usage.") All documents are made available at no charge to member-company distribution points. Document ordering is facilitated by the *Technology Transfer Product Catalog,* a publication with cross-referenced listings of available publications, products, and services.

Consortium Speakers Bureau

We make consortium technical staff members available to address member-company groups at member-company locations. In the past, the consortium has provided speakers for a variety of different meetings, ranging from company-wide symposia to project- and program-specific engineering team gatherings. Consortium speakers cover such topics as process improvement, software reuse, software productivity metrics, software acquisition management, development methods, domain analysis, and software modeling.

Workshops and Seminars

The consortium has discovered that technology workshops provide one of the best means of ensuring collaboration with its members during the initial stages of product development. Such workshops and seminars are offered at the consortium throughout the year; both tend to focus on a single technical issue. Workshops provide an open forum for attendees to conduct a substantive dialogue, contribute actual project experience to discussions of product plans and functionality, and allow participants to provide further direction to emerging consortium technologies. Although consortium speakers often give presentations during a workshop, group interchange remains the primary goal. Participants have the opportunity to learn from the experience of other professionals, while at the same time contributing to the understanding of all workshop participants. As a result, the sessions are highly interactive. Past workshop themes have included system modeling, reuse, software design, and technology transfer.

While seminars also have a single-issue orientation, they feature distinguished lectures from government, industry, and academia. Unlike workshops, these sessions have a strong educational flavor. Seminars may include speaker presentations, required reading, and classroom exercises.

Advisory Groups

Member company engineers have the opportunity to gain insights into advanced technology concepts and influence consortium work through participation in a variety of consortium advisory groups. In addition to the Technical Advisory Board (TAB) described earlier, Technology Advisory Group (TAG) members work with the consortium project teams offering advice and comments on current and future technology investigations. TAG members get new perspectives on consortium products and technology, plus they have the opportunity to see their opinions realized in more relevant and valued products. TAG participation requires commitment of time and periodic visits to the consortium's facility in Herndon, Virginia.

The TAGs also facilitate the transfer of methods, technology, and tool sets, thereby increasing the benefits of member company participation in the consortium. By working in the TAG, group members concentrate on issues pertinent to the successful introduction of innovations to their companies. Because they are collaborative by nature, advisory groups are an excellent means of building a partnership approach to technology transfer. Our workshops, seminars, and speakers-bureau initiatives build and maintain the partnership dialogue.

Technology support services provide the specific, detail-oriented institution and assistance necessary to deploy new software processes and methods. These services include the following items.

Pilot Projects. Pilot projects are collaborative activities designed to ensure that each technology works in the customer's environment. Involving extensive personal support of a customer's development efforts by highly-accomplished consortium technical staff, pilot projects have proven to be an excellent technology transfer mechanism. Usually applied on "live," real-time systems-development programs, pilot projects allow the early implementation of new technologies and provide the customer with another means of influencing the consortium's technical activities in critical areas.

Education and Training. The consortium offers its member companies education and training services from its headquarters facility in Herndon, Virginia, or, based on demand, at customer sites. The curriculum is designed to provide maximum benefit for the relatively short time a student spends in a training class. For instance, all courses feature reference-based instruction; teaching is often focused on proper use of actual documentation. This approach recognizes the inherent difficulty of retaining large amounts of detailed technical information introduced in a classroom situation. Rather than ask a student to absorb such details, reference-based training provides practitioners with a general understanding of product functionality and an in-depth knowledge of how to find execution details in guidebooks, reference manuals, and user guides. These documents are the consortium's primary training material, and each student returns to the member company with a complete set.

For tool sets and prototypes, classes also feature contextually reinforced training. Students use the tool sets to evaluate real-time software-development problems, such as the design of multiprocessor bus architecture. Operating in a "hands on" mode, they use the tool sets to perform a variety of typical software engineering tasks. Exercises might include creating and editing a top-level software design; evaluating reusable parts for possible incorporation in a design; and assessing whether a design meets system performance requirements. Detailed scheduling and enrollment information is provided to member-company transfer sites and board members and technical advisory board and groups members.

Consortium educational offerings have proved to be very successful transfer mechanisms. On a yearly basis we provide more than 250 days to training and attract more than a thousand attendees to our workshops, seminars, and training classes.

Consulting Support. The consortium offers a well-coordinated set of consulting services, designed to cut the time necessary for engineers to begin using our products. For instance, the consortium maintains a product support center and "hot-line" product support telephone service. Staffed by software professionals thoroughly trained in the use of our products, hot-line service is available on business days between 8 A.M. and 5 P.M. EST. Product support center personnel answer member company questions concerning the application of our processes, methods, and tool sets. Because these are technical staff members, many questions can be answered immediately.

Demonstration Center. Our methods and tools sets are showcased in the consortium's demonstration center, located at our facility in Herndon, Virginia. Here we offer briefings and demonstrations on a variety of topics, ranging from general overviews of the organization and its technical directions to interactive demonstration sessions with tool prototypes and products. Member company personnel schedule visits to the demonstration center on an as-needed basis; product demonstrations can also be held at member sites, based on demand.

Presentations and demonstrations focus on both short- and long-term topics. For instance, the facility offers introductions to the functionality and architecture of our Ada design method and modeling tool set—currently available to member-company engineers. Briefings are also available on subjects with a more conceptual perspective or longer lead-time orientation. These include topics such as specific tool prototypes under evaluation, software development methods, architectural issues, and future tool development plans.

Finally, the customer feedback program seeks to increase customer satisfaction and provide for continuous improvement in the development of technologies that are useful to engineers working with them on a day-to-day basis. In addition to the collaboration and partnership mechanisms described above, the consortium is putting in place feedback mechanisms, which include short survey forms aimed at eliciting comments on the value and usefulness of our products and to identify areas where improvement is needed. Questionnaires and in-depth interviews are also planned, all focused on furthering the transformation of the consortium into a "customer-driven" company.

This is perhaps the most important lesson we have learned in technology transfer: a consortium must be as committed, or even more commited, to satisfying its customers as are commercial organizations. The unique mission and size of the Software Productivity Consortium demand that we develop products that our members/customers truly need; aggressively promote and educate thousands of software engineers on their correct usage; and constantly gather and analyze feedback to further improve these technologies. Only through such an active transfer strategy can we continue to succeed.

In summary, the consortium is offering a cohesive set of services designed to support technology transfer. Table 11.1 cross-references these transfer initiatives with the "steps" to product adoption and customer satisfaction.

The nature of consortium membership provides member companies with equal opportunities but not necessarily equal value. The value of membership depends on a sponsoring company's sustained interest in the technical program. If the company loses interest, program technology loses value because it will not be applied. We seek to generate and maintain interest in our program through extensive, two-way communication with all of our customers.

A consortium is an organization owned by its customer base. As a result, when communication between the partners is inadequate, any consortium is apt to be pulled in several directions at once. In our case, we often run up against strong advocates of a short-term return on investment. But yielding to such impulses would diminish our strategic focus and radically change the nature of our quest. The consortium charter is to define a new software-development process model. Why? Because present day CASE tools offer small productivity gains. We are creating methods and building tools that implement our process model, the Evolutionary Spiral Process, and asking commercial vendors to follow suit through our Strategic Alliance Program. A strong overall communication program has helped us navigate in such currents.

We have learned, however, that communication miscues can make it difficult to maintain steady progress. Early on, for instance, we thought product demonstrations and presentations might be conducted on a multimember company scale. We have since seen that company interest levels vary in particular products and that interest in a new technology tends to converge at a conceptual level. As a result, we have had more success addressing broad software engineering themes in multiple company workshops.

While interest is nurtured by communication within the partnership, each member company's "inclination" to be an active consortium member seems to be driven by multiple factors, including internal priorities and other facets of the corporate culture. Inclination seems to show itself in other ways. For instance, the linkage between transfer points in the member companies and the consortium can be very secure and direct. We know this is the case because we respond to multiple requests for service: telephone consultation, marketing support, education and training, workshop participation, technical reports, and other items. We have instances, however, where member-company response is not as active as we would like. A key challenge of the consortium technology transfer program, then, is to influence all partners to use our technology and provide reasonable levels of feedback.

Our member-company assignee program has been conducted with mixed success. We had many expectations for this approach, including the use of returning assignees as consortium transfer agents. This was not always the case, however; experience indicates that the best place for resident assignees is in positions where they are involved in technology transfer activities. So, we have learned to place them on such projects and to maintain close contact with them once they return to their companies. In this way, they can be our most vocal champions within our member companies.

Organizational awareness also helps us gauge member inclination. Although the consortium charter specifies the use of consortium-developed tools and methods on real programs, many of our transfer sites are, necessarily, internal software technology research groups. We understand and support our customer's need to "kick the tires," and such research-oriented groups provide an important evaluative service. On the other hand, as the VCR example suggests, our nation's technology transfer shortcomings seem to roost around the hand-off between the research lab and the line project.

To effect his hand-off in the software arena, we continue to implement strategies that stimulate and sustain member-company interest while working to persuade the "gatekeepers" posted within our transfer sites to disseminate consortium technology to engineers working on live projects. To this end, we have packaged our products with as much commercial flavor as a constrained marketing budget will allow and continue to establish alliances with commercial vendors for CASE tool support for our technologies. Another strategy is to build prototypes and products in smaller functional increments, but deliver them on a more frequent basis. This approach, coupled with interactive programs such as technical advisory and user groups, has made it easier for companies to digest our innovations and gives them a stronger sense of ownership. Most importantly, the approach works: as mentioned earlier, we now can count dozens of instances where our customers have identified the usage of our products as "key" to the success of a project, division, or the customer organization itself.

We will continue to work with the software industry to facilitate the intergrability of consortium technology with popular commercial-off-the-shelf-tools. Our Strategic Alliance Program, begun just this year, has already resulted in the development and marketing of CASE tools which support our Ada design method and our modeling tool set; other alliances are in the works. By positioning our methods and tool sets with proven marketplace products, we think our members will be attracted by the potential for synergy and seek opportunities to exploit their installed base of software. So rather than compete with commercial vendors, we encourage them. Our hope is that much of the necessary tool set automation can be provided by third parties, leaving us the task of method and process refinement.

In addition to interest and inclination, successful technology transfer will depend on the competitive ability of our member companies. In fact, this observation goes to the heart of the consortia model. As the innovation builder and supplier, we certainly strive to make each of our members more productive and, as a result, more competitive. In a sense, we are not only in the innovation business, we are in the opportunity business also. After all, innovation, especially in the software industry, is an ephemeral quantity. To the extent our members seize the opportunity and reap competitive advantage from our technology, we are delighted. If any member does not recognize the opportunity offered—either through poor communication or in different support—we will have failed not only as technology transfer agents, but as providers of beneficial technologies.

Partnership with our members and customers remains the key to our continued success. As I hope this chapter has described, the Software Productivity Consortium is building a partnership with major aerospace, defense, and electronics companies through a multifaceted technology transfer program.

References

Chand, Donald R. and Sridhar A. Raghavan. "Diffusing Software Engineering Methods," *IEEE Software*, July 1989.

Cooper, Robert G. "The New Product Process: A Decision Guide for Management," *Journal of Marketing Management*, Spring 1988.

Corcoran, Elizabeth. "Technology Transfer—Research Consortiums Look for Ways to Work with Backers," *Scientific American*, May 1989.

Scacchi, Walt and James Babcock. "Understanding Software Technology Transfer," Technical Report STP-309-87, Microelectronics and Computer Technology Corporation, Austin, Texas, October 1987.

Yourdon, Edward. "A Game Plan for Technology Transfer," *IEEE Software*, 1987.

Part IV

Overcoming Barriers to Technology Transfer

12

Building a Technology Transfer Infrastructure[1]

Raymond W. Smilor
David V. Gibson

Spurred on by increasing international competition, the rising costs of advanced research, the need to leverage scarce scientific and technical talent, and the desire to share the risk associated with technology generation and commercialization, high-technology firms have begun to band together in cooperative research activities (Fausfeld and Haklisch, 1985; Dimancescu and Botkin, 1986, Bopp, 1988). As a result, R&D consortia have become a global phenomenon. Management in these firms now faces the intriguing paradox of competition and cooperation. To compete more effectively in international markets, they must find innovative ways to cooperate (Dimancescu, 1987; Inman, 1987, 1988; Borys and Jemison, 1988; Noyce, 1989; Kozmetsky, 1989; Smilor, Gibson and Avery, 1989).

Since the passage of the National Cooperation Research Act in October 1984, as of December 1989 over 150 R&D consortia have filed with the U.S. Department of Justice. Some of the more prominent U.S. consortia are the Microelectronics and Computer Technology Corporation (MCC) and SEMATECH in Austin, Texas; the Software Engineering Institute in Pittsburgh; the National Center for Manufacturing Science in Ann Arbor, Michigan; the Semiconductor Research Corporation in Research Triangle Park, North Carolina; the Biotechnology Research and Development Corporation in Peoria, Illinois; the Software Productivity Consortium in Herndon, Virginia; and the Center for Advanced Television Studies in Boston.

These consortia cover a range of emerging technologies, including telecommunications, microelectronics, semiconductor manufacturing, biotechnology, software engineering, transportation and superconductivity (see Figure 12.1). They are composed of 1,157 business, government, and academic organizations. Data on 902 of the business organizations shows that 843 are U.S. companies, and 42 are foreign firms. The balance are state and federal government agencies, U.S. and foreign universities, and other consortia.[2] Table 12.1 shows the U.S. firms involved in six or more consortia.

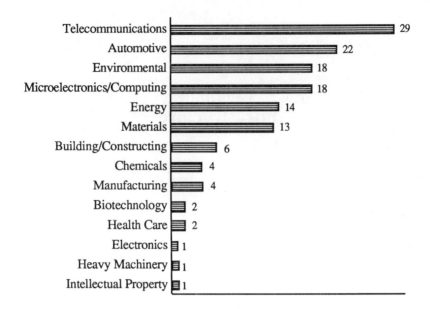

Figure 12.1. U.S. R&D consortia by industry type.

R&D consortia have also been developing in Europe and Japan. European consortia include ESPRIT in information technology, RACE in advanced communications, BRITE in advanced materials and manufacturing, JESSI in semiconductors, and JFIT in information technologies. Japanese firms have formed the Very Large Scale Integration (VLSI) Consortium for high-capacity memory chips ICOT for the fifth-generation computer, artificial intelligence, and parallel processing and the Tsukuba Research Consortium composed of eight "core companies" for joint research in such areas as magnetic levitation and advanced computer modeling of bacterial photosynthetic organs.

New Organizational Form

R&D consortia represent a new organizational form, and they pose unique management challenges (Peck, 1986; Levinson and Moran, 1987; Evan and Olk, 1988; Souder and Nassar, 1988; Stotesbery, 1988). Consortia are often composed of companies (i.e., shareholders) that seek mutually beneficial cooperative research while remaining fierce competitors in the marketplace. They are often composed of personnel from radically different corporate cultures. In addition, shareholder members present different managerial priorities, policies and procedures. As a result, companies often participate in a consortia for different and sometimes conflicting reasons. Within any single consortia, different research programs, utilizing different research methodologies, are commonly being pursued simultaneously. Consortia management and researchers are thus separated from shareholder organizations by a variety of professional, technological, strategic and cultural barriers (Murphy, 1987; Gibson, Rogers, and Smilor, 1988; Smilor, Gibson, and Avery, 1989).

Table 12.1. Companies Involved in Six or More Consortia

Belcore (telecommunications)	22
Digital Equipment Co. (electronic computing equipment)	9
Texas Instruments (semiconductors and related devices)	8
Rockwell Corporation (aircraft)	8
Hewlett-Packard (electronic computing equipment)	8
Ford (motor vehicles and car bodies)	8
Honeywell (radio and TV communications equipment)	7
Harris Corporation (radio and RV communications equipment)	7
General Motors (motor vehicles and car bodies)	7
EXXON (crude petroleum and natural gas)	7
Amoco Corporation (petroleum refining)	7
Shell Development Company (petroleum refining)	6
Mobil R&D Corporation (crude petroleum)	6
IBM (electronic computing equipment)	6
General Electric Company (turbines and turbine generating sets)	6
E. I. DuPont de Nemours (organic fibers, noncellutosic)	6

While consortia vary in organizational structure, technological emphasis, funding mechanisms, and personnel make-up, they all share one abiding issue which relates to their purpose for being formed—the transfer of technology to their member companies in an efficient, timely manner (Devine et al., 1987; Inman, 1987, 1988; Kozmetsky, 1988a, 1988b, Noyce, 1989; Pinkston, 1989; Rogers, 1989).

Technology transfer is the application of knowledge (Weick, 1988; Segman, 1989). It involves any geographical shift of technology (ideas as well as physical products): person-to-person, group-to-group, or organization-to-organization. There are two fundamental aspects to the transfer process. First, the technology must be created or discovered. Second, it must be expeditiously transferred to the appropriate receptor. This second aspect is proving to be at least as challenging and certainly more controversial than the first (Inman, 1987; Pinkston, 1989).

Methodology

An applied management model for technology transfer has been developed from a year-long study of the MCC (Microelectronics and Computer Technology Corporation) a major, for-profit, U.S. R&D consortium. In-depth interviews were conducted with managers and scientists at the consortium and with MCC's shareholder representatives. The interviews were supplemented by archival and observation data. Based on these data a survey was developed and was distributed throughout the consortium in January 1989.

Of a total sample of 430 possible respondents, 147 completed and returned the survey for a response rate of 34 percent. The questionnaire focused on the following aspects of technology transfer at MCC: effectiveness of technology transfer at the consortium; effectiveness of various methods for technology transfer; importance of various factors in facilitating the technology transfer process; importance of barriers to technology transfer at both the consortium and the shareholder companies; and agreement on ways that the consortium could improve the technology transfer process.

The following figures present the mean scores and standard deviations on the survey questions by two categories of MCC employees: direct hires and shareholder representatives/employees. MCC direct hires come from industry, university, and government backgrounds and are full-time employees of the consortium. Shareholder representatives are employees of the shareholder companies are but are housed at MCC for a period of two to three years.

Effectiveness of MCC's Technology Transfer

Figure 12.2 shows that MCC direct hires and shareholder representatives/employees share a high level of agreement on the following statements: technology transfer is a major requirement for the success of MCC (5.6, 5.7); technology transfer should be a top priority at MCC (5.4, 5.4); and there have been problems with the technology transfer process at MCC (5.0, 4.9). Both groups share the least agreement on following three statements: technology transfer is largely the responsibility of the shareholder companies (3.4, 3.6); MCC is effective in transferring technology to the shareholder companies (3.4, 3.3); and technology transfer is largely the responsibility of MCC (3.1, 3.2).

There is disparity of agreement between MCC direct hires and shareholder representatives/employees concerning the statement that technology transfer from MCC is a major requirement of the shareholder companies. While both groups agree with this statement, the shareholder representatives/employees more strongly agree than do MCC direct hires (5.4 to 4.6). Each group indicates that they have a better understanding of the importance of technology transfer than their counterparts. MCC direct hires more strongly agree that there is a good understanding of the importance of technology transfer at MCC (4.7 to 4.4). At the same time, shareholder representatives/employees agree more strongly that there is a good understanding of the importance of technology transfer at the shareholder companies (4.1 to 3.5). Interestingly, both groups agree that there is less understanding of the importance of technology transfer at the shareholder

Building a Technology Transfer Infrastructure • 133

companies than at MCC. MCC direct hires also more strongly agree than do shareholder representatives/employees that technology transfer is now a top priority at MCC (4.6 to 4.3).

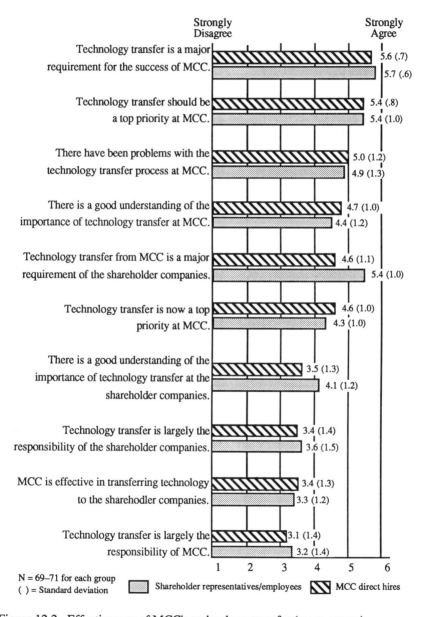

Figure 12.2. Effectiveness of MCC's technology transfer (mean scores).

Effectiveness of Methods for Technology Transfer

Figure 12.3 shows that MCC direct hires and shareholder representatives/ employees share similar attitudes on the effectiveness of most technology transfer methods as well as on their ranking from very effective to very ineffective.

There is some difference of opinion on five of the methods. MCC direct hires consider tutorials (4.5 to 4.1), demonstrations at MCC (4.4 to 4.0), and refereed journal articles (3.3 to 2.7) more effective than shareholder representatives/ employees. On the other hand, shareholder representatives/ employees consider themselves (4.8 to 4.4) and proprietary technical reports (3.8 to 3.4) more effective than MCC direct hires.

Importance of Factors Facilitating Technology Transfer

Figure 12.4 shows that MCC direct hires and shareholder representatives/ employees share similar opinions on their ranking from very important to very unimportant the factors facilitating technology transfer.

There is a notable difference of opinion on two of the factors. MCC direct hires consider knowing who to contact more important than shareholder representatives/employees (5.5 to 5.0). Shareholder representatives/employees consider a service/customer-oriented attitude more important than MCC direct hires (4.8 to 4.6).

Importance of Barriers to Technology Transfer

Figure 12.5 shows that MCC direct hires and shareholder representatives/ employees share similar perceptions of the importance of most of the barriers to technology transfer at MCC and at the shareholder companies.

A number of differences, however, are noteworthy. MCC direct hires consider the following barriers more important than shareholder representatives/employees: "not invented here" syndrome at the shareholder company (4.8 to 4.3); no clear definition of technology transfer at the shareholder company (4.8 to 4.3); not knowing who to contact at the shareholder company (4.8 to 4.4). Shareholder representatives/employees, on the other hand, consider the following barriers more important than MCC direct hires: "not invented here" syndrome at MCC (4.0 to 3.6) and secrecy at MCC (3.0 to 2.6).

Ways to Improve Technology Transfer

Figure 12.6 shows that MCC direct hires agree more than shareholder representatives/employees that the following ways would improve the technology transfer process: involve shareholder researchers more with MCC (5.3 to 4.9), increase awareness of importance of technology transfer within shareholder companies (5.1 to 4.8), have shareholders communicate to MCC more (4.9 to 4.5), provide more incentives to shareholder personnel for transferring technology (4.8

Building a Technology Transfer Infrastructure • 135

to 4.1), and provide more incentives to MCC and personnel for transferring technology (4.3 to 3.8).

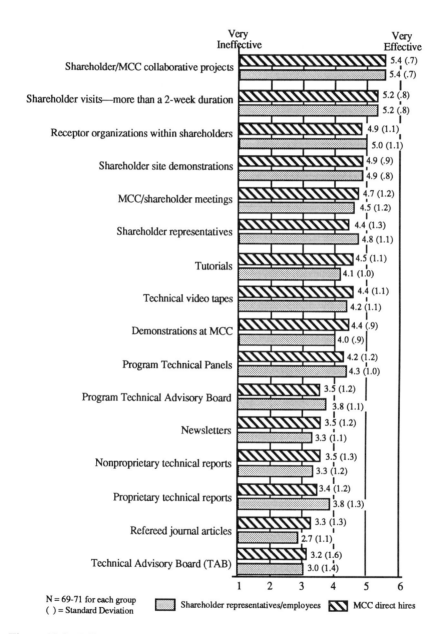

Figure 12.3. Effectiveness of methods for technology transfer (mean scores).

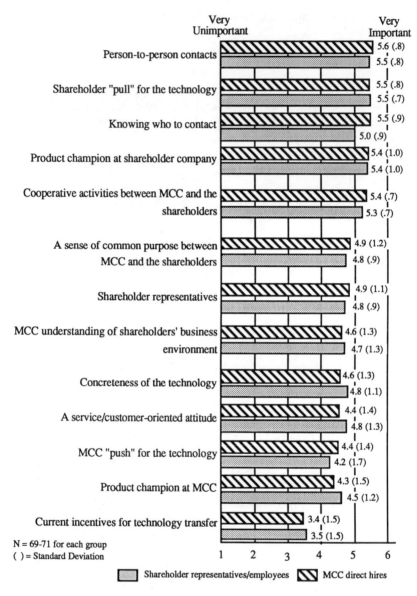

Figure 12.4. Importance of factors facilitating technology transfer (mean scores).

Shareholder representatives/employees agree more than MCC direct hires that the following ways would improve the technology transfer process: share success stories of technology transfer among program areas (4.4 to 4.2), increase awareness of importance of technology transfer within MCC (4.3 to 4.1), and involve shareholder marketing and product planning personnel more with MCC (4.0 to 3.4). Both groups indicate less agreement that setting up a technology transfer office for each research program, establishing a technology transfer committee for each MCC program, and setting up a technology transfer office at MCC would improve the technology transfer process, with MCC direct hires favoring these ways more than shareholder representatives/employees.

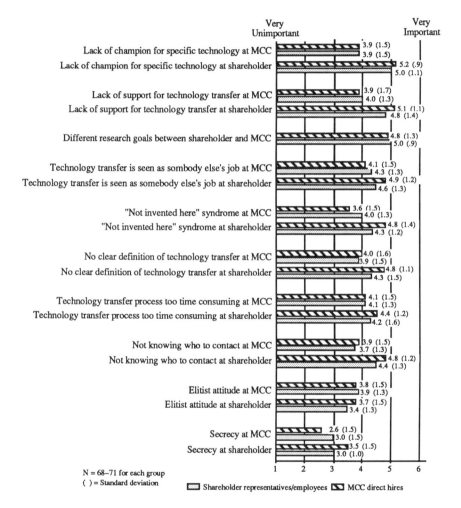

Figure 12.5. Importance of barriers to technology transfer (mean scores).

Both groups disagree that establishing a research program in technology transfer at MCC and using outside consultants would improve the process, with shareholder representatives/employees indicating the strongest disagreement with establishing a research program in technology transfer at MCC.

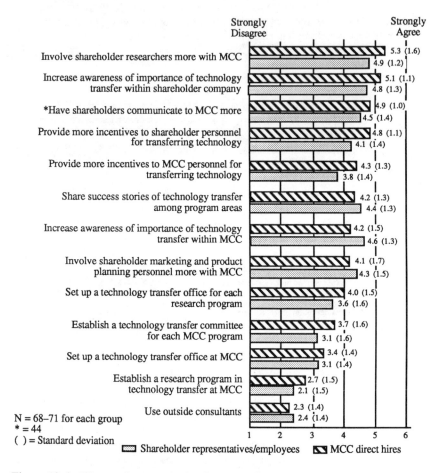

Figure 12.6. Ways to improve technology transfer.

A New Management Model for Technology Transfer

Based on the survey, interview, and archival data, four key variables emerge as especially critical in the technology transfer process: communication, motivation, distance, and technological equivocality. Management can influence, direct, and monitor each of these variables. As shown in Figure 12.7, there is an important relationship among these variables which can be ranked from low to high.

Communication

Communication between the technology transmitter and receptor involves both passive and active links (Avery, 1989). Communication interactivity is closely related to information-carrying capacity, which refers to the degree to which a medium is able to efficiently and accurately relay task-relevant information (Daft and Lengel, 1984; Huber and Daft, 1987; Sitkin, et. al., 1989) and media richness (Daft and Lengel, 1986). Passive links have a broad sweep and are usually media based. Included in the passive category are research reports, journal articles, computer tapes, and video tapes. Such media-based linkages are considered best for rapidly communicating the same message at the same time to a widely dispersed audience at a relatively low cost. Furthermore, greater care can be taken to package or produce a quality message.

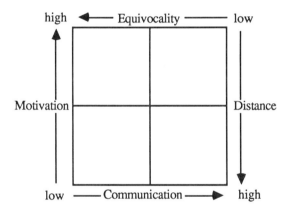

Figure 12.7. Technology transfer variables.

Such media-based forms of communication are generally nonthreatening to the technology transmitter in that the researcher can "stay at home" in his base location. It can be professionally threatening to go into the field (e.g., a shareholder company) to actively try to transfer one's research to potential receptors. However, with such media-based transfer attempts, the transmitter is often unaware of whether and how the receptors receive and utilize the transferred technology. Media-based communication may go out from the consortium, but no feedback may return from the shareholders.

Of the hundreds or thousands of possible shareholder receptors to broad-based, consortium-generated, technology transfer attempts, each receptor interprets the media-based communicated message in his/her own way, and each has his/her own agenda of research priorities, time allocations, and resource commitments concerning his/her championing of the communicated technology. The hope is that in selected instances the media-based message will hit the receptor at just the right moment, and the receptor will be motivated to take the time and resources to pursue a more active linkage with the consortium.

Active links are direct, person-to-person interactions. They may range from teleconferences to ad hoc teams and on-site demonstrations. The benefits of active links center on the fact that they encourage interpersonal communication in terms of fast, focused feedback, i.e., the consortium researcher learns from the potential user and vice versa. Such face-to-face communication, however, is more costly in terms of time commitments and other resources of the consortium and the shareholder receptors. Furthermore, they require the consortium and shareholder personnel to have the ability, desire, and time to communicate across organizational boundaries to effect technology transfer.

The fewer and more passive the communication links, the less likely the chance that technology will be successfully transferred. The higher or more active the communication links, the more likely the chance of technology transfer.

Distance

The second variable, distance, involves both geographical and cultural proximity or separation (Pinkston, 1989). Geographical distance within the consortium or between the consortium and shareholder companies involves the physical proximity of those engaged in research and information activities. Distances based on building, floor, office, interregion or interstate separation can slow or inhibit the technology process (Rogers and Kincaid, 1982; Hatch, 1987). The lower or closer the geographical distance, the more likely the chance of successful technology transfer. Management can lessen the physical distance of those engaged in research activities by promoting more active and direct communication links and by co-locating technology transmitters and receptors by various ad hoc and formal means.

Cultural differences within the consortium or between the consortium and the shareholders loom as a more important dimension of distance than geographical separation (Albrecht and Ropp, 1984; Borys and Jemison, 1988; Smilor, Gibson, and Avery, 1989). The diversity of corporate cultures among the many shareholders at the consortium pose significant managerial challenges in dealing with the distance variable. Each shareholder brings his/her own values, attitudes, and ways of doing things to the consortium. The higher or wider the cultural distance between the consortium and the shareholder company, the more difficult it is to transfer the technology. The lower the distance, i.e., the more the consortium researchers and personnel understand the values, attitudes and ways of doing things in the shareholder company, the greater the chance of technology transfer.

Equivocality

Equivocality refers to the level of concreteness of the technology to be transferred (Weick, 1988; Pinkston, 1989; Avery, 1989). Technology that is low in equivocality is fairly easy to understand and demonstrate and is unambiguous. That is, the meanings the technology conveys to different individuals is highly similar, i.e., objective. The lower the level of equivocality, the more likely that the technology will be transferred. On the other hand, highly equivocal technology is harder to understand, more difficult to demonstrate, and more ambiguous in its potential applications. The higher the equivocality of the technology, the more difficult it is to educate perspective users on the value/applications of that technology.

Motivation

Motivation involves incentives for and recognition of technology transfer. The motivation for participating in the technology transfer process can range from low to high for both technology generators and receptors. Motivation varies with importance of technology transfer in the culture of an organization, the criteria by which the individual is evaluated, and the rewards established for those who engaged in technology transfer activity (Dornbush and Scott, 1975).

Both technology transmitters and receptors can legitimately ask, "What's in it for me?" The greater the degree of variety of incentives, rewards, and recognitions, the higher the motivation for those engaged in the process. Managerial systems that promote active communication, that foster understanding of and appreciation for different cultures, that encourage cooperative activities to increase proximity, and that lower the equivocality of technology will heighten motivation and increase the chances of successful transfer.

Technology Transfer Grid

The technology transfer grid, as shown in Figure 12.8, describes four organizational situations affecting technology transfer. These situations reflect various combinations of the variables in the technology transfer process.

Technology transfer is "dead in the water" when there is low communication, low motivation, high distance, and high equivocality. In this situation, technology transfer cannot occur because transmitters and receivers do not talk with one another, because there are neither incentives nor recognition for those involved in the process, because there are wide geographical and cultural distances, and because the technology is ambiguous and the applications uncertain. The technology may be developed, but it is neither accepted nor commercialized.

At the other end of the spectrum is the "grand slam" situation. In this case, all elements are right for technology transfer. There is high communication, high motivation, low distance, and low equivocality. In other words, because of highly interactive communication processes, because of a variety of incentives and recognitions, because of geographical and cultural proximity, and because the technology is unambiguous and its application understood, successful technology transfer occurs.

The two other situations described in the technology transfer grid are the "black hole" and the "long shot." While there are a variety of combinations of variables in each situation, each is initially characterized by two positive and two negative variables.

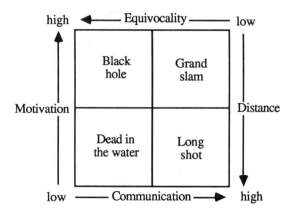

Figure 12.8. Technology transfer grid.

"Black hole" describes the situation in which there is high motivation and low distance combined with low communication and high equivocally. In other words, while these may be rewards for technology transfer as well as geographical and cultural closeness, communication processes tend to be passive and indirect, and the technology is ambiguous and uncertain in its applications. In this situation, information may be passed from transmitters to receivers, but there is little or no feedback. The black hole concept thus refers to the difficulty of transmitters in getting adequate feedback for passive transfers of technology. In short, when technology transfer initiatives such as technical reports, videos, and computer tapes are sent to shareholder companies, they may be "swallowed up" without the consortium scientists knowing who, if anybody, has taken note of, or even whether the "technology" has been passed on to the most appropriate receptor. While there are various combinations of the variables in this situation, they tend to result in this "black hole" phenomenon, in which, for one reason or another, transfer is perceived to be a one-way process.

"Long shot" describes a situation in which there is a high communication and low equivocality combined with low motivation and high distance. In other words, while there may be active and direct communication processes in conjunction with technology that is unambiguous with clear applications, rewards are minimal and the geographical and cultural gaps are wide. In this situation, people may interact and understand the technology, but they experience difficulty in selling or marketing ideas with unmotivated participants across geographical and cultural barriers. In short, when technology transfer initiatives such as demonstrations, on-site visits, and person-to-person contacts occur, transfer is a long shot because of the prevalence of attitudes such as a not invented here syndrome, a perception that technology transfer is somebody else's responsibility, and the feeling that it is too time-consuming a job on one side or the other or both.

Management Action

Given these factors, shareholder and consortia management can take action to develop an infrastructure that is supportive of and conducive to technology transfer (Creighton et al., 1985). As shown in Figure 12.9, these actions are grouped according to the four key variables identified in the Technology transfer grid. The actions from the perspectives of the shareholder, consortium management, research program areas, and individual scientists are thus designed to increase communication and motivation and to reduce distance and equivocality.

Communication Actions

Recommendations to improve the effectiveness of communication are designed to increase the number and range of active mechanisms and to disseminate more broadly and effectively passive mechanisms of communication. Shareholders should clearly identify and give authority to a receptor organization and/or persons in the shareholder company to monitor, receive, and appropriately disseminate, at the shareholder organization, technology developed at the consortium; provide a regular column in internal shareholder newsletters to present important developments, personnel, and technology transfer activities to increase awareness of the consortium throughout the shareholder organization; and assign a liaison to the consortium who is visible and highly regarded within the shareholder company to help champion the technology transfer process.

Consortium management should set up a technology transfer office to monitor, document, and publicize technology transfer activities; develop a mythology of technology transfer by publicizing and disseminating success stories of technology transfer; and establish a quarterly technology transfer newsletter to highlight technology transfer activities and to recognize those involved in them.

Technical program areas should disseminate passive communication mechanisms to a broader base of constituents within shareholder companies to encourage serendipitous linking of interests and applications; develop a target market of possible technology receptors at the shareholder companies through a database of selected production, marketing, sales, and R&D personnel at each shareholder organization; conduct more informal meetings with shareholder personnel to encourage interpersonal relationships; increase teleconferencing activities; and increase teaming activities to expand individual ownership of the technology.

Individual consortium researchers should know who to contact in the shareholder company to facilitate technology transfer; write up technology transfer activity for publication in newsletters; and develop personal networks within the consortium and among shareholder companies to increase the number and diversity of possible technology transfer contacts.

144 • Technology Transfer in Consortia and Strategic Alliances

Action Areas	Communication *Active/Passive Mechanisms*	Distance *Geographical/Cultural Proximity*	Equivocality *Concreteness of Technology*	Motivation *Incentives/Rewards/Recognition*

Shareholder/
Top Management
	• Identify receptor organization
	• Identify person to contact
	• Provide regular consortium column in internal newsletter
	• Assign liaison
		• Expand number, quality, and diversity of people interacting with consortia/alliance partners
		• Place researchers in the consortium/alliance
		• Involve product/marketing personnel
		• Identify product champions
			• Clarify expectations for research activities
			• Clarify usability criteria technologies
			• Encourage collaborative projects/activities
				• Provide situational incentives/rewards
				  - Teaming opportunities
				  - Collaborative projects
				  - Training opportunities
				  - Funds for pet projects
				  - On-site visits
				  - Longer-term exchanges
				  - Visiting lecturers

Consortium/
Alliance Management
	• Develop mythology of technology management
	• Establish technology management column in newsletter
	• Identify key person in partner companies
		• Document success stories
		• Have consortium personnel act as consultants to shareholders
		• Develop training/education programs on technology management
		• Set up corporate culture workshops
			• Require consortium/alliance people/programs to have technology transfer objectives
			• Conduct research trade shows
			• Develop education/training programs on selling ideas
			• Encourage demonstrations at consortium and on-site
				• Provide financial incentives/rewards
				  - Bonuses
				  - Pay raises
				  - Licensing/royalty arrangements
				  - Honorariums

Program/Project Areas	• Disseminate passive mechanisms more broadly
	• Develop a target market of technology receptors at shareholders
	• Conduct more informal meetings with shareholder personnel
	• Increase teleconferencing activity
	• Increase teaming activities
		• Conduct interactive workshops at consortium/alliance
		• Arrange visits for extended periods at both consortium and alliance
		• Identify technology champion
		• Develop service/customer-oriented attitude
		• Expand open-house activities
			• Develop collaborative projects to facilitate interactive assessments
			• Customize the transfer process
			• Educate partners to the benefits of the technology
			• Conduct demonstrations/tutorials
				• Provide recognition
				  - Newsletters
				  - Dissemination of success stories
				  - Video tapes
				  - Annual awards banquet
				  - Monthly consortium/alliance awards

Individual	• Know who to contact • Write up technology management activity for publication in newsletters • Develop personal networks	• Visit partners early and frequently • Participate in cooperative activities	• Participate in technology transfer programs • Participate in demonstrations/tutorials • Participate in collaborative projects	• Include technology management efforts in performance appraisals

Figure 12.9. Technology transfer infrastructure

Distance Actions

Recommendations to decrease the distance between the consortium and the research program areas on the one hand and the shareholder companies on the other are designed to increase geographical and cultural proximity. Shareholder management should expand the number, quality, and diversity of people interacting with the consortium to increase mutual understanding of values, attitudes of ways of doing things in the consortium, and the shareholder organization; place researchers in the consortium for extended periods of time; involve product/marketing personnel in consortium research activities; hold technology transfer seminars to bring together consortia and shareholder personnel at the shareholder organization; identify product champions at the shareholder organization; and begin a distinguished lecture series to bring in key consortium managers and scientists to discuss the culture, research, and different approaches to technology transfer at the consortium.

Consortium management should begin a distinguished lecture series to bring in key shareholder managers to discuss the culture, approach to innovation, and role of technology transfer at the shareholder organization; document success stories of technology transfer; have consortium personnel act as consultants to the shareholder companies on selected projects; develop training/education programs on technology transfer and commercialization; and set up corporate culture workshops to provide consortium personnel with an understanding of the culture of each of the shareholder organizations.

Technical program areas should conduct interactive workshops at the shareholder companies and consortium to more effectively link program area research activities with shareholder application requirements; arrange visits for extended periods for program area personnel to visit the shareholder companies and for shareholder personnel to visit the program area; identify consortium technology champions who can effectively interact with a range of personnel at the shareholder company; develop a service/customer-oriented attitude by educating program area personnel to the needs, requirements, and preferences of the shareholders; and expand open-house activities to tap a broad base of expertise across the consortium program areas and shareholder companies.

Individual researchers should visit shareholder companies early and frequently throughout the research activity and participate in cooperative activities within the consortium and among shareholders.

Equivocality Actions

Recommendations to lower the equivocality of the technology are designed to make the technology more concrete, more understandable to the user, and less ambiguous.

Shareholder managements should clarify expectations for research activities so that consortium personnel have a better understanding of what the shareholder expects to get from its involvement with the consortium; clarify the usability criteria for consortium technologies so that consortium personnel understand specific shareholder needs and requirements; set up internal meetings for presentation by consortium personnel to provide direct feedback on research activities; and encourage collaborative projects to facilitate sharing of information and research results.

Consortium management should require consortium people/programs to have technology transfer objectives; conduct research trade shows consortiumwide, where appropriate, to make potential users more aware of developing technologies; develop education/training programs on selling ideas; and encourage demonstrations at the consortium and on-site to make the technology more understandable to potential users of the technology.

Research program areas should establish early linkages with shareholders to understandable usability criteria; develop collaborative projects with a diversity of shareholder personnel to facilitate interactive assessments of the research activity; customize the technology transfer process to fit the specific policies, procedures, and requirements of each shareholder; educate shareholders to the potential benefits of the technology as well as to its specifications and features; utilize program technical panels to provide shareholder direction on research activities; and conduct demonstrations/tutorials at the consortium and at shareholder organizations.

Individual researchers should participate in conferences that focus on understanding and facilitating the technology transfer process; participate in demonstrations/tutorials to champion one's own research; and participate in collaborative projects to increase awareness of shareholder needs and expectations.

Motivation Actions

Recommendations to heighten motivation focus on providing incentives, rewards, and recognition for those involved in transferring technology. These recommendations apply across action areas for the shareholder, consortium management, and program areas.

Situational incentives and rewards for those involved in technology transfer may include teaming opportunities with well-known and highly respected personnel, innovative collaborative projects that allow for interaction with a diversity of individuals and groups, training opportunities that expand an individual's knowledge and expertise, a special allocation of funds for a pet project, time and resources for on-site visits, participation in longer-term exchanges to experience the implementation of research activities, and selection as a visiting lecturer.

Financial incentives and rewards may include more traditional monetary compensations such as bonuses and pay raises, special licensing/royalty arrangements for transferred technology, and honorariums for particularly noteworthy achievements in technology transfer.

Recognition may include featuring individuals and groups in newsletters, in documentation of success stories, and in video tapes describing technology transfer activities. Recognition may also include participation in a major, annual technology transfer awards banquet and selection for a monthly consortium/shareholder technology transfer award. In addition, technology transfer efforts should be part of one's performance appraisal.

Conclusions

Technology transfer has become a key issue in determining the success or failure of R&D consortia. Those involved in managing consortia can accelerate the technology transfer process by taking the initiative to develop an infrastructure that focuses on increasing communication and motivation while decreasing distance and equivocality. In doing so, R&D consortia are more likely to contribute to making the shareholder companies more competitive in the international marketplace.

Notes

1. Portions of this article have appeared in R. W. Smilor and D. V. Gibson, "Technology Transfer in Multi-Organization Environments," *IEEE Transactions on Engineering Management,* 1991, and "Accelerating Technology Transfer in R&D Consortia," *Research/Technology Management,* 1991.

2. Consortia database, IC^2 Institute, the University of Texas at Austin.

References

Albrecht, T. L. and V. A. Ropp. "Communication About Innovation in Networks of Three U.S. Organizations." *Journal of Communication,* Summer 1984, pp. 78–91.

Avery, C. "Technology Transfer Issues in U.S. Consortia." Doctoral Dissertation, College of Communication, The University of Texas at Austin, 1989.

Bopp, G. R., ed. *Federal Lab Technology Transfer: Issues and Policies.* New York: Praeger, 1988.

Borys, B. and D. B. Jemison. "Hybrid Arrangements as Strategic Alliances: Theoretical Issues in Organizational Combinations." *Academy of Management Review,* 1988.

Creighton, J. W., J. A. Jolly, and T. A. Buckles. "The Manager's Role in Technology Transfer." *Journal of Technology Transfer,* vol. 10, no. 1, Fall 1985, p. 67.

Daft, R. L. and R. H. Lengel. "Information Richness: A New Approach to Manager Information Processing and Organization Design," ed. B. Staw and L. Cummings, *Research in Organizational Behavior,* vol. 5, 1984, Greenwich, Conn: JAI Press.

_____. "Organizational Information Requirements, Media Richness and Structural Design." *Management Science,* vol. 32, no. 5, 1986, pp. 554–71.

Devine, M. D., T. E. James, Jr., and I. T. Adams. "Government Support Industry-Research Centers: Issues for Successful Technology Transfer." *Journal of Technology Transfer,* vol. 12, no. 1, Fall 1987, pp. 27–38.

Dimancescu, D. "The Competitive Challenge: Can America's R&D Consortia Respond?" "Managing the Knowledge Asset into the 21st Century: Focus on Research Consortia," proceedings from a conference at Purdue University, 1987, pp. 8–13.

_____ and J. Botkin. *The New Alliances: America's R&D Consortia Respond?* Cambridge, Mass.: Ballinger Publishing, 1986.

Dornbusch, S. M. and W. R. Scott. *Evaluation and the Exercise of Authority.* San Francisco: Jossey-Bass, 1975.

Evan, W. M. and P. Olk. "R&D Consortia: A New Organizational Form." Technical report, The Wharton School of Business, 1988.

Fausfeld, H. I. and C. S. Haklisch. "Cooperative R&D for Competitors." *Harvard Business Review,* November-December 1985

Gibson, D., E. Rogers, and R. W. Smilor. "The Importance of Multi-Constituency Communication to Research Consortia: The Case of the MCC." Paper presented to the Symposium on Science Communication, the Annenberg School of Communication, University of Southern California, December 1988.

Hatch, M. "Physical Barriers, Task Characteristics, and Interaction Activity in Research and Development Firms." *Administrative Science Quarterly,* 32, 1987, pp. 387–99.

Huber, G. H. and R. L. Daft. "Information Environments," ed. F. Jablin, L. Putman, K. Roberts and L. Porter *Handbook of Organizational Communication.* Beverly Hills, Calif.: Sage Publications, 1987.

Inman, B. R. "Commercializing Technology and U.S. Competitiveness." *High Technology Marketing Review,* 1(2), 1987, pp. 83–89.

_____. The founding President, CEO and Chairman of the Board, MCC. Interview, 1988.

Kozmetsky, G. "The Challenge of Technology Innovation in the Coming Economy." Presented to the 13th Annual Symposium on Technology Transfer Society, Portland, Oregon, July 1, 1988a.

_____. "Commercializing Technologies: The Next Steps," ed. G. R. Bopp, *Federal Lab Technology Transfer: Issues and Policies,* New York: Praeger, 1988b, pp. 171–82.

_____. "Tomorrow's Transformational Managers," ed. K. D. Walters, *Entrepreneurial Management: New Technology and New Market Development,* Cambridge, Mass.: Ballinger Publishing Company, 1989, pp. 171–76.

Levinson, N. S. and D. Moran. "R&D Management and Organizational Coupling." *IEEE Transactions on Engineering Management,* 34. no. 1, February 1987, pp. 28–35.

Murphy, W. J., III. "Cooperative Action to Achieve Competitive Strategic Objectives: A Study of the Microelectronics and Computer Technology Corporation." Dissertation. Cambridge, Mass.: Harvard School of Business, 1987.

Noyce, W. M. The founding and current President and CEO of SEMATECH. Presentation to the Graduate School of Business, The University of Texas at Austin, 1989.

Peck, M. J. "Joint R&D: The Case of Microelectronics and Computer Technology Corporation." *Research Policy,* 15, 1986, pp. 219–31.

Pinkston, J. T. "Technology Transfer: Issues for Consortia," ed. K. D. Walters, *Entrepreneurial Management: New Technology and New Market Development,* Cambridge, Mass.: Ballinger Publishing Company, 1989, pp. 143–49.

Rogers, D. M. A. "Entrepreneurial Approaches to Accelerate Technology Commercialization" ed. K. D. Walters, *Entrepreneurial Management: New Technology and New Market Development,* Cambridge, Mass.: Ballinger Publishing Company, 1989, pp. 3–15.

Rogers, E. M. and D. L. Kincaid. *Communications Networks: A New Paradigm for Research.* New York: The Free Press, 1982.

Segman, R. Editor of the *Journal of Technology Transfer.* Presentation at the Roundtable on Technology Transfer, Lago Vita, Texas, July 28, 1989.

Sitkin, S. B., K M. Sutcliffe, and J. R. Barrios-Choplin. "Determinants of Communication Media Choice in Organizations." Working paper, Graduate School of Business, The University of Texas at Austin, April 1989.

Smilor, R. W., D. Gibson, and C. Avery. "R&D Consortia and Technology Transfer: Initial Lessons from MCC." *Journal of Technology Transfer,* vol. 14, no. 2, Spring 1989, pp. 11–22.

Souder, W. E. and S. Nassar. "Advantages and Disadvantages of R&D Consortia." *Research–Technology Management,* 1988.

Stotesbery, W. D. "Improving Competitiveness Through Collaborative Research," ed. G. R. Bopp, *Federal Lab Technology Transfer: Issues and Policies,* New York: Praeger, 1988, pp. 87–96.

Weick, K. "Technology as Equivoque: Sense-making in New Technologies." Paper presented at the Conference on Technology, Carnegie Mellon University, August 28–30, 1988.

13

Technology Transfer from Academia: Prescription for Success and Failure

John F. Hesselberth

This chapter discusses the important subject of industry/academia interactions and, more specifically, technology transfer from universities to industry. While I write from the perspective that improving technology transfer from academia to industry through the mechanism of a consortium is a desirable goal, Du Pont's experiences with consortia have not been very positive, with a few exceptions. In the present chapter I will present an analysis of factors important to success as contrasted with a prescription for failure.

To put Du Pont's experiences and observations on consortia into a more understandable context, I will briefly examine the total spectrum of interactions between industry and academia. By far the simplest of these interactions is an unrestricted grant made by industry to a university or a specific school or department within the university. Nothing specific is expected in return and no legal agreements are involved. Industry makes these grants as a way of contributing to the education process and, at least in Du Pont, they are focused on schools who do the best job of turning out graduates we want to hire. Du Pont gave $14 million in unrestricted grants to universities in 1989. Du Pont also gives specific grants to an academic department to work in a field of interest to Du Pont. Often there is no contract involved. In these cases the company usually gets the right to use the technology developed royalty free, but does not get exclusive rights to the technology.

Consulting is perhaps the most common of the various Du Pont university interactions and also is quite straightforward. All that is usually needed is a simple nondisclosure agreement and a contract specifying compensation and commitment of time. If the arrangement does not prove beneficial to both sides, it is easy to modify or discontinue. The next order of complexity is represented by a research contract. Again, this type of agreement is normally between two parties, although in this case the individual professor is normally replaced by the university as the academic representative to the agreement. As we will see, this adds to the complexity; however, on the whole, these arrangements are still fairly straightforward. Adding together specific grants and research contracts, Du Pont funded over $15 million of these types of work in 1989.

On the surface consortia may also seem simple and straightforward. After all, for a consortium already in existence or being organized by a university, you either join or you do not. Joining is usually a simple matter of paying a few thousand dollars a year. Beyond that, nothing is required. In theory a consortium would seem to be an ideal concept. It offers a way to leverage the technology the university develops or, stated more precisely, a way to disseminate technology more broadly and quickly than other mechanisms seem to allow. These statements I have just made, however, are a shallow explanation indeed. Consortia are, by their very nature, complex arrangements which require a heavy commitment of time and resources on both the part of universities and industry if, and that is a big "IF," anything useful is to come out of them.

This is not to say that consortia are not worth the effort. They can be worthwhile. However it takes more thoughtful planning and more effort than we often seem to be willing to devote to making them useful and that is the main point I would like to emphasize here: making consortia useful requires thoughtful planning and a lot of hard work on the parts of both industry and academia.

Research Contracts and Consortia

In the following pages I will offer some data that summarize Du Pont's experiences about a specific success in research contracts and consortia, and some observations on what contributes to such successes. I will also describe some of the characteristics Du Pont observes in the less successful interactions, and offer a prescription for failure. In making these observations, I purposefully focus on the extremes—the very best and the very worst—in order to draw the most specific lessons. This chapter ends with a list of lessons that can dramatically increase the value of university/industry interactions, whether they be research contracts or consortia.

I know what I write in the following pages will be controversial; I intend it to be. It is offered in the spirit of starting a dialogue on this critical subject with the objective being a step-change improvement in technology transfer from academia to industry. Achieving a step-change improvement would represent major progress toward improving our national competitiveness. Today, too much technology is developed that never gets used and that is a tragic waste of a national resource. Technology transfer is the primary criteria by which I judge the success or failure of industry/academia research interactions. If technology transfer does not occur, the interaction cannot be judged to be successful.

Du Pont's Experience with Consortia

Du Pont periodically polls the research and development community across the corporation to determine how much research money is being spent outside the company and to obtain an informal assessment of the value of that research. The question on value is simple: Whether the benefit received has exceeded expectations, has met expectations, or has been lower than expectations. Du Pont divides its poll into several classifications, four of which are pertinent here today. We ask about (1) research contracts between Du Pont and other industrial corporations, (2) research contracts between Du Pont and universities, (3) consortia that are formed between several industrial corporations, and (4) consortia with a university focus.

As indicated in Table 13.1, sample size varies widely. Du Pont has had far more experience with research contracts than with consortia. However, when asked whether or not the benefits received exceeded expectations, it is evident that in spite of the small number of data points, the results are dramatically different between research contracts and consortia.

Table 13.1. Du Pont's Experience with Research Contracts and Consortia

Type of Research Agreement	Sample Size	Percent Exceeding Expectations
Contract—Company	129	29.5
Contract—University	183	22.4
Consortium—Company	8	12.5
Consortium—University	22	4.3

I do not agree that there is any significant difference between multicompany consortia and university-sponsored consortia; however, I do make a case that the difference between research contracts and consortia is disappointingly large. I have also found that expectations for getting something useful from consortia are lower than for research contracts. People tend to join consortia so as not to be left out of the club, rather than expecting specific useful results to be forthcoming.

The Positive Side

As implied in Table 13.1, 4.3 percent of 22 equals 1. The one university consortium exceeding expectations is the Center for Process Analytical Chemistry (CPAC) at the University of Washington at Seattle. Du Pont has found CPAC to represent the best characteristics of university/industrial consortia today. A number of other university consortia are meeting Du Pont's expectations, and the line between meeting and exceeding expectations is admittedly fuzzy and subjective;

however, CPAC is certainly one of the best, and it can well serve as an example to study in a little more detail.

CPAC started as a National Science Foundation (NSF) "Center" and was about number 20 in a progression of 50 centers. The NSF gave CPAC a total of $500 million in start-up funding over a five-year period, and CPAC "graduated" to financial independence. CPAC still competes for and gets research funding from the NSF and others, but the start-up funding has been stopped. At last count CPAC had 48 sponsors and the number was growing. One of the issues addressed at CPAC's last annual meeting was whether or not they were becoming so crowded they were losing some of their effectiveness. CPAC decided not to limit membership, but they are still struggling with the problems of success.

CPAC has done several things right which have enabled them to turn into such a valued consortium in a relatively short period of time.

- First, they found an area that was very important to industry but was being virtually ignored by academia. In other words, they found an unmet need. The need was strong and was put into terms that the industries who were solicited for membership could relate to. The "offering" was unique.
- Second, the University of Washington provided for energetic, capable, full-time administration. The administrator must have a lot of enthusiasm for what he or she is doing because there will be plenty of opportunity to get discouraged, particularly in the early years. That the administrator must be capable goes without saying, but more specifically, I believe the administrator must have a good balance of technical skills in the area of interest, administrative skills, and interpersonal skills. The challenge of forming a team effort around a specific area of technology with traditional industrial competitors being the team members is not a simple challenge.
- Third, CPAC developed specific research objectives and programs that meet the needs of their industrial partners. Too often, consortia twist existing research efforts to make it sound like they are offering something that is needed when, in reality, they are merely dressing up existing programs in a new package. One problem Du Pont has with regard to managing our relationship with CPAC is that their programs are perceived to be of such high value that we have scientists and engineers standing in line to be on their technical monitoring committees. Often, with other consortia, we have to twist arms to find people willing to serve in such capacities.

A Prescription for Failure

It's time to look at the other side of the question. Why are so many of the consortia Du Pont encounters less than positive experiences. The bottom line, of course, is that Du Pont feels it does not get technology that is valuable. In other words, effective technology transfer does not occur. But let's look at the question from a different angle and ask (1) why was the consortium formed or, in other words, what is its basis for existing, and (2) how is it managed from both academia's and industry's perspective. In developing a prescription for failure I may sound somewhat cynical, but, in any case, it is my perception that these things happen more frequently than desired.

The first part of my prescription deals with the basis on which the consortium is founded. The first and perhaps most assured way to achieve failure is to start a consortium as a way to attract new funding to an existing area of research. In order to make it sound more attractive, the founders wrap it in the trendiest, most political nomenclature: for example, advanced materials or composites. They draw in existing staff who are doing research that is somewhat related, with full assurances that they will have to make minimal, if any, changes to their ongoing research programs.

A second, and almost equally effective, way to assure failure is to join the lemmings on their march to the sea. There are certain fads that get started in the technological community that seem to turn into the "glamorous" area to work. Everyone wants a piece of the action even when there is only a need for a few. For at least a short period of time, money seems to flow freely to these areas. No one wants to coordinate efforts to avoid overlap; everyone wants to be the focus of the effort. The latest example of this is in the area of environmental protection. I tried to find out how many university-sponsored consortia there are in this general area. I did not succeed. A new one seems to spring up at least once a week. There is no apparent coordination between those already formed and no discernable attempt to do so. Some are sponsored by governmental entities, either federal or state, and some deliberately are trying to stay nonaligned. Clearly, environmental protection is an extremely important area and one that the nation needs to address aggressively. But can we afford to do it in this "join-the-bandwagon" type of approach? I do not believe so. I predict that after an initial blush of success most of these consortia will flounder and eventually die for lack of industrial interest. The ones that succeed will find a niche, having significant interest and will then strive to become truly world class in that niche even if it means bringing in new staff to accomplish this leadership.

Industry can play its role in assuring failure also. Industry may have a need and encourage the "local" university to initiate a consortium in the area because it is geographically convenient to do so. The result can be a halfhearted attempt on the part of the university to accommodate their industrial neighbor to the ultimate detriment of both. Another factor is the way industry decides to join a consortium. Too often, the decision is made based on preserving personal or organizational relationships rather than on an objective assessment of needs vs. the offering.

The fourth part of my prescription for failure starts with seemingly harmless questions, like "What should we do with the laboratory space that just became available?" and "How can we obtain funds to operate it?"

I think it is important to contrast the previous prescriptions for failure with the following: forming a consortium by combining a recognized industrial or societal need for knowledge and know-how that is not being met and a scientific or engineering skill base that has the technical capabilities to meet the need. This is the best way to avoid failure and have a fighting chance at success. That is one of the things CPAC did so well.

The second part of my prescription for failure deals with the management of the consortium and here my prescription is straightforward and easy to carry out.

The first element is to pick a part-time manager who would really rather be doing research. Convince the person that this is a job that should not take more than ten percent of his or her time. The same goes for industry partners. They can assure failure by asking a disinterested person who already has technical challenges that totally consume him or her to be a technical adviser or monitor in their "spare" time. Working out the details of a research program that must provide benefit to multiple parties is a time-consuming, difficult task. It requires people who are not only interested in assuring success, but also those who can see the big picture versus getting bogged down in detail, and who can find creative ways around the inevitable impasses that will develop between the interested parties.

A second element for failure is to write global-sounding objectives so everyone will be able to see themselves as benefiting. This will assure that, in the end, no one is satisfied. It is the "one-size-fits-all" gimmick. It would be closer to the truth to say "one size fits no one." Rather, objectives must be clear and specific, even if that means some of the partners decide the consortium really does not meet their needs. At least the remaining partners will have a better chance of being able to utilize the results of the work.

A third element of a consortium management that will assure failure is to staff the consortium with new, inexperienced people who have not developed a strong research reputation. This element is also applicable to both industry and academia. Industry must put some of our very best people into the arrangement. Yes, there is room for participation by less experienced people, but a foundation of demonstrated research capability must be present at the table.

Summary

In summary, the first factor to focus on to assure success is finding a real need in industry that is not being met by others. Ideas must be put on the table, studied, discussed with potential partners, modified, and discussed again and again before the concept for the consortium is finalized. Second, the skills and interests of the university must match the need. This is equally important to defining the need, because it will ultimately determine the commitment and the value of the university effort. Third, energetic and capable people must be identified to manage the effort, and they must have enough time to be able to do it well. Fourth, the objectives must be stated specifically, programs must be designed that really address the objectives, and mechanisms must be put in place to monitor progress. Finally, both industry and academia must be willing to devote significant time and effort to making the partnership work.

14

Partnerships with Industry and the Oak Ridge National Laboratory

Warren D. Siemens

Productive R&D partnerships should create win-win situations. Each partner must bring something to the partnership that is needed by the other partners, and each partner must be able to leave the partnership with more than it contributed. However, if it were that simple we would have more successful R&D partnerships in this country.

Partnerships of various kinds, and consortia in particular, are being propounded as the answer to regaining our international competitiveness. Indeed, partnerships may be a very effective vehicle to rally the interest, create the momentum, leverage the dollars, and marshal the technological resources to a competitiveness challenge, in lieu of any national strategy for a technology. In effect, the establishment of a major U.S. industry R&D consortium becomes a grass-roots effort to establish a national agenda for a technology area. If it fully draws on all the necessary resources, it will have gotten the participation of industry, government, and universities. However, developing the cooperation and participation of the key parties in a partnership is a monumental task, requiring strong leadership. Many considerations would dictate against effective partnerships in this country, such as the way we conduct business and the adversary roles of industry and government.

Some of the factors, in Oak Ridge National Laboratory's (ORNL) experience, that seem to most influence whether an R&D partnership can be successful are discussed in the following pages. These factors were developed as a result of conducting a postmortem on a consortium initiated by ORNL that never gained sufficient participation by industry. The consortium was called "CAMDEC," and its demise has been analyzed in "CAMDEC, an Industry R&D Consortium: Lessons Learned" (Siemens and Cadotte, forthcoming). Examples are also given of industry partnerships with the ORNL that have been initiated as a means of making the technologies, facilities, and expertise more available to industry and to accelerate the commercialization of government-funded technology developments.

Factors Influencing the Success of R&D Partnerships

A broad spectrum of issues and factors that may contribute to the success or failure of an R&D partnership include cultural influences, industry environment, company priorities and practices, and consortia design. Even though there is no single model to follow, understanding these issues and factors may improve the odds of success.

Cultural Influences

Two of the more important cultural factors are beliefs regarding competitor cooperation and the need for short-term profits. These cultural traits are intangible and tend to change slowly over time. Although corporate beliefs are changing to be more tolerant of cooperation among competitors, many obstacles still remain that limit the full potential of such partnerships.

Rugged individualism. Consortia are based on cooperation between competitive firms. This requirement runs counter to an entrenched, fundamental American ideal which Edward Deming calls "rugged individualism." It is our desire to succeed independently. That is, Americans take pride in individualism. It sets us apart from each other and allows individual creativity to thrive. To limit this creativity, as do the Japanese, may inhibit America's drive for ingenuity.

Short-term focus. Shareholder emphasis on profitability has forced American corporations to focus on short-term investment options. This short-term (less than two years) strategy is not compatible with the long-term (greater than five years) strategy of a consortium.

Industrial Environment

Business climate can greatly affect the willingness of an industry to participate in a consortium. Important industrial factors include the character of the market environment, the nature of competition, and external threats.

Character of the market. Market uncertainty, size, and maturity are factors that determine whether a consortium can be successfully implemented. For many technology applications there are major market uncertainties that affect the risk in joining a consortium. Uncertainty regarding the size of the market, the speed at which it will develop, the overall health of the industry, the true advantage of the proposed technology, the need for industry standards, and the possibility that another technological innovation could in turn displace the proposed technology add to the risk of a joint venture.

Market maturity may also be a driving force behind consortia. Many consortia have been formed to revitalize the market and create new opportunities for participating firms. One example is the International Partners in Glass Research.

This consortium is based on a common desire by industry participants to stimulate growth and meet the threat from non-glass bottling technology.

Nature of the competition. The intensity of the competition in an industry is determined by various factors, such as the scope of distribution (local, regional, national, or international), the number of competitive firms (domestic and foreign), and the basis for their differential advantage (protection of trade secrets and intellectual property).

Consortia seem to be more successful when industry participants do not compete "head to head." Regulated industries like utilities distribute their services regionally and are more likely to form a successful consortium than more highly competitive industries. For instance, the Gas Research Institute and the Electric Power Research Institute are supported by utility companies, which generally serve regional markets.

At the other end of the competitive spectrum are industries that are highly competitive. If there are many competitors, or they have a history of aggressive market tactics to steal market-share points, or each firm's differential advantage is based upon know-how, trade secrets, and internally developed technologies, it is very difficult to form consortia. In many firms, proprietary processing techniques are their advantage over the competition. Therefore, unless trade secrets can be managed successfully, a firm's risk from losing valuable information may outweigh the benefits gained from joining a consortium.

External threats. There is an important exception where highly competitive industries have been able to work together. It is usually the case when the industry perceives a common, outside "threat." If several members of the industry believe that their very existence is threatened by foreign competition, a leading domestic competitor, or a new technology that replaces their own, then they may be able to put aside individual differences and work together. In many cases, they have their backs against the wall and are forced to cooperate with each other for long-term survival.

Company Priorities and Practices

The internal politics of a corporation have proved to be factors influencing a firm's willingness to join consortia and its utilization of new technology. A number of these factors are discussed below.

Not invented here. Too often, companies distrust consortium-developed technology because it was not developed in their own research laboratories. A firm's mid-level managers and scientists may also resent that the "glamour" research is being given to consortia, while they conduct the more mundane applications research.

Scarce resources. A related issue is the competition for scarce resources within a company. When consortia are funded, company support is usually drawn from existing budgets. This causes internal conflicts because company researchers would prefer to see those limited corporate R&D dollars channeled to their own programs rather than to the consortium.

Market-pull philosophy. It has often been said that U.S. industry should not wait for the market to develop before investing in advanced technology but should invest money, make a product, and create a market in order to be competitive. The current bias in industry seems to be that the only technologies worth pursuing are the ones for which there is a clear market with known customers. This philosophy does not imply that they will not invest in longer-term technology development efforts, but they have to have one or more customers in mind and would want to work with those customers during the development period.

Technical gatekeepers. From our experience, a senior member of the research staff always plays a key role in the decision to join an R&D consortium, even if the entry point was with a senior executive. The key technical individual is usually assigned the responsibility of carefully evaluating the consortium and making a recommendation to top management. They are essentially gatekeepers in the approval process. Many executives feel it would not be wise to override the recommendation of their senior research staff and force them to accept a consortium without some compelling reason to justify it.

Consortia Design

The structure of a consortium is obviously very important and is probably the easiest to blame for a consortium's lack of success. Some of the more critical factors include the nature of the technology agenda, the uncertainty of its benefits, a level playing field, the cost of membership, leadership from a strong champion, and funding from a sugar daddy.

Technology agenda. The research and development agenda should be set by the industry. Ideally, the agenda should reflect the strategic goals of participating member firms and it should determine the research to be conducted with quantitative goals and milestones for the technology. Since developing a consortium agenda involves strategic issues of importance to the whole industry, those decisions should not be left solely to the R&D managers. Top management and the business planners must be involved. The agenda must specify what technological advances are to be targeted. "Leaps" in technology are usually needed to gain the interest of corporations because "small steps" can usually be funded within an individual firm's R&D facilities.

Benefit uncertainty. An obvious problem with consortia is the uncertainty of achieving the technological breakthroughs required to meet their objectives, particularly, if the leaps in technology are not credible. In many cases, a consortium is formed to tackle difficult technologies which no single firm has the resources or know-how to solve. One can never be certain that the technical problems will be solved, that the technology can be scaled up to production levels, or that it will come on line in a timely manner to help the industry. By its very nature, a consortium effort will be risky.

Level playing field. When different firms participate in a consortium, their relative risk and return is usually not equivalent. Individual companies are at different points on the experience curve. A company with less experience in a technology has the potential for greater benefit from a consortium because of the opportunity to catch up with its competitors, whereas a company that is more advanced in a technology perceives high risks and small returns because it has less to gain from joining the consortium. Efforts to create a level playing field often make it difficult to agree on a technology agenda such that the risks and benefits are comparable for all players.

Costs of membership. Annual membership fees for consortia range from tens of thousands to millions of dollars. To these costs, members must also add complementary costs which follow from participation in the consortium, such as travel expenses, salaries of participating corporate employees, opportunity costs of alternative investments, and the cost of a shadow program to adopt the technology when it becomes available.

Champion. A champion is an individual or organization who is dedicated to the success and completion of a consortium and provides the leadership role. The champion is critical to the successful initiation and follow-through of a cooperative venture. An example of a lead individual is William Norris, CEO (retired) of Control Data, who mobilized the CEOs of the computer industry to fund MCC.

Sugar daddy. The "sugar daddy," although not unlike the champion, is an organization, public or private, that funds a major part of the research. A recent Office of Technology Assessment report indicates that 42 percent of the funding for computer and microelectronics consortia and 77 percent of the funding for advanced materials consortia came from organizations other than private industry. This type of funding leverages a firm's investment and decreases the risk of its investment. This type of funding from a "sugar daddy" will often entice competing firms to negotiate, agree, and cooperate on a research agenda.

Industry Partnerships Initiated at ORNL

Partnerships with industry are being promoted by the Office of Technology Applications at Martin Marietta Energy Systems, Inc., as a means to bridge the gap between R&D and commercialization. Through the utilization and application of technologies, facilities, and expertise at the Oak Ridge National Laboratory and the other Department of Energy facilities in Oak Ridge that are managed and operated by Martin Marietta, technological resources are being made more available to the private sector.

While the Office of Technology Applications has received considerable recognition for its aggressive licensing program, it believes that in the long run greater impact on the U.S. economy will be realized if industry can effectively access and acquire the technologies at the national laboratories through partnership arrangements with industry. We believe that through such partnerships, government-sponsored R&D can be more rapidly and broadly commercialized.

A broad range of partnership activities with industry have been initiated at ORNL. Some of them involve direct partnerships with industry, while others involve an intermediary organization that organizes the industry participation.

Tennessee Center for Research and Development

In 1985, the Tennessee Center for Research and Development (TCRD), a not-for-profit corporation, was established on the technology corridor between Oak Ridge and Knoxville for the express purpose of creating economic value from the strong science and technology bases that exist in the region at such institutions as the Oak Ridge National Laboratory, the University of Tennessee, and the Tennessee Valley Authority (TVA). Martin Marietta Energy Systems was instrumental in getting TCRD initiated and currently participates on its board of directors. The purpose of the center is to bridge the gap between R&D and commercialization by supporting market-driven applications development. It allows R&D organizations in the Oak Ridge–Knoxville region to "mature" their technologies beyond what may be appropriate to their respective missions. TCRD is a halfway house to develop commercial applications of government-developed technologies. TCRD's activities may be funded by a variety of sources, including federal and state agencies, industry associations, individual companies, and private investors. It draws on the R&D resources and capabilities of the organizations in the region to accomplish its objectives through various consulting and contractual arrangements.

Power Electronics Applications Center. One of the first major achievements of TCRD was the establishment of the Power Electronics Applications Center. Developing the proposal for the center provided the first opportunity for three organizations—Energy Systems, TVA, and the University of Tennessee—to aggressively work together on initiating a regional activity. The purpose of the center is to regain the competitive position of the power electronics industry in this country through the development, demonstration and transfer of power electronics technologies for U.S. companies. Initial, multiyear funding was provided by the Electric Power Research Institute. However, R&D partnerships are currently being formed and funded by interested organizations to conduct specific power electronics developments for such applications as adjustable speed drives, power-line conditioners, and uninterruptable power supplies. The objective is to develop high-voltage and high-current electronic devices and systems which provide more efficient electricity end-use management. Development areas include devices and components, circuits and controls, industrial electrotechnology systems, power conversion and conditioning systems, and power quality. As industries become more productive users of electricity, their product costs become more competitive, and they become healthier electric utility customers, which has a moderating effect on electricity rates. Thus, all economic sectors ultimately benefit from advances in power electronics.

Thermomechanical Model Software Development Center. This center was initiated by Energy Systems under the umbrella of TCRD. Twelve sponsoring companies, listed in Table 14.1, are funding the development of a user friendly and intelligent software system for more easily accessing a highly complex set of finite element analysis models. These models, which analyze thermomechanical stresses in refractory systems, such as furnace linings, were developed by MIT for ORNL over an eight-year period. In their current configuration they are too complex for industrial design engineers to use, requiring persons with considerable background and experience in thermomechanical modeling, thermomechanical behavior, mechanics, and finite element analysis.

Table 14.1. Companies Participating in the Thermomechanical Model Software Development Center

Refractory Companies
- C-E Refractories
- Institut de Recherches de la Sidérurgie
- Norton Company
- Standard Oil Engineered Materials
- Radex

Coal Gasification Process Developers
- AMOCO
- Shell Development
- Texaco

Steel Companies
- ARMCO
- ASEA Industrial Systems
- LTV Steel
- National Steel Corporation

In addition, preparation of the input files requires considerable time, and interpretation of the results requires considerable background and experience with the model. The consortium development will allow design engineers to input the necessary design parameters and interpret the analysis results without being an expert in these complex models. Such a system will be valuable to those concerned with refractory behavior in coal gasification systems as well as in other systems, which include high-temperature refractory applications such as blast furnaces in the steel industry.

The project is being cosponsored by the DOE Fossil Energy Program. Industry is cost sharing this three-year project on a 50–50 basis. Each industrial partner will receive, as part of its consortium fee, two systems installed at their facilities. This industrial consortium effort is augmenting a national laboratory technology to significantly increase its commercial attractiveness and utility.

The High Temperature Superconductivity Pilot Center

A new mechanism for developing and transferring technology to the private sector was implemented at the Oak Ridge National Laboratory during FY 1989. The ORNL High Temperature Superconductivity Pilot Center was authorized to form partnerships with U.S. companies to develop commercial applications for high-temperature superconducting materials and devices.

This pilot program in public-private cooperation was initiated by the Department of Energy (DOE) in 1988 to make the resources of the national laboratories available to U.S. companies to speed up the process of moving from research to commercial products. In addition to Oak Ridge, two other centers were sited at Argonne, Illinois, and Los Alamos, New Mexico. The three laboratories were selected because of their current involvement in superconductivity research and previous experience in technology transfer and industry collaboration.

To encourage industrial participation in the pilot centers, DOE provided some new tools. With input from the laboratories and industry, DOE developed a "model" cooperative agreement containing flexible contractual arrangements and new attractive provisions for patent and data rights. The research agenda for each agreement is developed jointly by the laboratory and its industrial partner. All projects are cost-shared, and, in most cases, no funds change hands. Each party performs its respective tasks, and the results are shared between the partners. The agreements allow and encourage extensive communication and interaction between the laboratory and industrial researchers.

The ORNL HTSC Pilot Center is funded through DOE's Office of Energy Storage and Distribution. During its first year of operation, the ORNL pilot center signed agreements with ten firms (Table 14.2) totaling $2.5 million; $1 million of DOE funds leveraged with $1.5 million of private industry funding. The scope of work represented by these projects covers a broad range of research at the laboratory spanning various divisions at ORNL, including metals and ceramics, solid state, energy, chemistry, and engineering technology.

Three companies associated with the pilot center program have also signed agreements with the High Temperature Materials Laboratory, a user facility that allows their scientists to use its special research facilities for materials characterization. Five of the user facilities at ORNL have capabilities of particular interest for work on high temperature superconductors.

The High Temperature Superconductivity Pilot Center provides a mechanism for one-on-one relationships between a national laboratory and the private sector. A separate technical agenda is negotiated with each client company, although joint ventures are encouraged and, in fact, are occurring. Success in this pilot partnership program could serve as a model for developing industry relationships in other key technology areas, and thus provide a way to combine the best of basic and applied research with the best capabilities in manufacturing and commercialization, a combination that many experts believe is essential if U.S. competitiveness and technological leadership is to be restored.

Table 14.2. ORNL, HTSC Pilot Center Cooperative Agreements

Company	Status	Technology Area
General Electric	Signed 12/88	Thallium HTSc material processing
Westinghouse Electric/ Univ. of New Mexico	Signed 4/89	Powder scale-up and wire fabrication
Corning	Signed 5/89	Deposition on flexible ceramic substrates
American Superconductor	Signed 6/89	Fabrication of wire and tape
Consultec Scientific	Signed 7/89	Deposition target device
Du Pont	Signed 7/89	Thin film devices and bulk applications
American Magnetics/ Georgia Tech	Signed 8/89	Characterization of multifilament conductors
Textron Specialty Materials	Signed 8/89	Deposition of conductors
GE/SUNY-Buffalo	Signed 10/89	Laser deposition of conductors
Superconductivity, Inc.	Signed 10/89	Magnetic energy storage

Collaborative Agreement on Gelcasting R&D

ORNL holds several patents applications for a technique called "gelcasting," which is used to form complex shapes. The process involves mixing a dense "slip" or suspension of a ceramic powder in a solution of a polymerizable organic compound. A polymerization initiator is added, and the slip is cast in metal or glass molds. The ceramic is then heated to polymerize (gel) the organic compound, after which the component is removed from the mold, dried, and the binder is driven out with further application of heat. Then, if desired, it can be machined to final dimensions. Finally, the part is sintered. The advantages of this technology are

(1) items can be molded to virtually the exact desired final dimensions, and (2) and final machining can be done on an intermediate-phase product so that no machining is required on the final sintered part. Since sintered ceramics are generally extremely difficult to machine (requiring diamond grinding) this process saves considerable expense in production of such items as alumina substrates for microcircuits and insulators, and ceramic components for mechanical applications.

Technologists at ORNL believe that gelcasting can be applied to silicon nitride ceramics, which have potential applications in heat engine components, such as turbocharger rotors. Here the advantage of near-net-shape casting would be very significant because of the geometric complexity of the parts. Silicon nitride could also be used for high-temperature components of such engines, perhaps bearings, or even the engine block. Such materials are expected to find considerable use within the next five years in advanced heat engines in order to reduce weight, increase operating temperature, reduce friction, and increase component lifetime and reliability. Considerable progress has been made in the development of ceramic components for those applications by the DOE Office of Transportation Systems programs. However, these programs have identified several critical areas requiring additional research before widespread use of ceramic components is likely to occur. While ORNL has preeminent skills in the gelcasting technique itself, it has little experience in fabrication of silicon nitride ceramics and no capabilities for hot isostatic pressing, a key process step for such applications.

To support its mission for heat engine research, Energy Systems determined that it would be beneficial to enter into one or more collaborative agreements with industry to investigate the feasibility and benefits of gelcasting for such applications. Since the laboratory's resources are limited in this area, it could not enter into agreements with all parties that might be interested in pursuing such collaborative R&D. Therefore, Energy Systems published an announcement in the *Commerce Business Daily* in August 1989 inviting industry to participate in a no-funds-exchanged collaborative agreement in gelcasting. It indicates that the industrial participant shall provide compounded powders for ORNL to form into test shapes using the gelcasting process. The parts shall then be returned to the industrial participant for densification by hot isostatic pressing. Responsibility for characterization of the resulting test parts will be shared by ORNL and the industrial participant.

Responses from vendors have been received and are currently being evaluated. It is expected that work will begin shortly after the first of 1990. ORNL scientists are optimistic that the synergism from this approach will provide exciting results for both DOE and commercial applications of gelcasting.

Optics MODIL

The Strategic Defense Initiative (SDI) Organization of the Department of Defense is establishing MODILs (Manufacturing Operations Development and Integration Laboratories) to accelerate the development of new technologies that are essential to the execution of the SDI producibility, affordability, and supportability strategy. MODILs link the laboratory to the factory through a knowledge and skills transfer to the industrial base.

The first MODIL to be established is in the area of survivable optics at the Oak Ridge National Laboratory. The Optics MODIL is to search for existing emerging and enabling technologies to solve manufacturing problems, develop where necessary state-of-the-art fabrication and metrology technologies, and establish a productivity and validation test bed to demonstrate and baseline prototype hardware for fabrication of high-precision optics systems.

A key premise in the Optics MODILs development is that perceived risk is the primary impediment to the adoption of a new or different fabrication method. Thus, program goals lean heavily on impartial technology evaluation and information distribution, communicated in terms that are quickly grasped by the business community.

The Optics MODIL will transfer the technology to the private sector through participation in specific process development programs, subcontracting demonstration projects, encouraging industry to assign engineers to work in the test bed, licensing the government-sponsored technology for commercial applications, and participation in information exchange meetings. Initially, industry participation will be funded through subcontracts to develop the necessary SDI technology capabilities in industry. Eventually, industry will be expected to cost share in meeting the manufacturing goals. The Optics MODIL will be promoting communication between optic designers, users, and manufacturers and between industry, universities, and government.

The strategic goals of the Optics MODIL are to reduce the cost of mirrors to 50 percent of today's R&D cost by 1992 and to 10 percent by 1995; to reduce lead time from the current 9 to 28 months to 3 to 6 months by 1992; to meet all performance requirements; and to provide alternative materials for the product.

Partnership with SEMATECH

In recent years the ORNL Fusion Energy Division and Solid State Division have developed expertise in gas plasma containment and diagnostics. Such systems have applications in the commercial world for fabrication of semiconductor devices. In November 1989, SEMATECH, the U.S. consortium formed to recover world leadership in semiconductor manufacturing, signed a contract with ORNL for the development of an advanced system for etching semiconductors using control techniques developed at the laboratory. SEMATECH is a consortium of 14 U.S. semiconductor manufacturers, with an annual budget of $200 million. Half of the funding comes from industry, or member firms, and the other half from the federal government through the Defense Advanced Research Projects Agency.

In a effort to help rebuild the semiconductor manufacturing infrastructure, SEMATECH will work with ORNL to develop Electron Cyclotron Resonance (ECR) plasma etch manufacturing technology. Several experimental ECR etch systems will be developed, evaluated, and optimized, and then SEMATECH will select the best configuration. The new technology will then be transferred to a U.S. semiconductor tool manufacturer who will incorporate the configuration into a production etch tool.

The technology that is the subject of this agreement relates to methods for manufacturing semiconductors with extremely narrow circuit paths. Present technology permits path widths in the range of three one hundred thousandths of an inch (0.8 microns). With the new technology it may be possible to reduce this by 40 percent or more. Also, because of the expected improvements in uniformity of the etching process over a larger area, the system should lower manufacturing costs and improve process quality. Ultimately this will result in much more compact and powerful computer and electronic products at lower prices.

The agreement with ORNL is part of SEMATECH's on-going equipment improvement program established to help improve or develop new manufacturing equipment and materials for the semiconductor industry. "Every time we sign a contract for the development of new technology and equipment for the semiconductor industry, we bring the United States a little closer to regaining world leadership in the semiconductor manufacturing industry. It's important for the U.S. to become proficient again in manufacturing," said Dr. Robert N. Noyce, the late president and CEO of SEMATECH. The contract with SEMATECH is part of the University and National Laboratory Program, which has identified leading scientists and educators at institutions around the country to form a network of "centers of excellence" to work under research grants from SEMATECH.

A challenge for ORNL will be performing the work on an accelerated timetable that meets SEMATECH's needs. Semiconductor technology is changing very rapidly, and there is a very narrow window of opportunity to create and catch the next generation production systems. ORNL intends to help SEMATECH do that with this project and is already looking forward to avenues for further cooperation with SEMATECH in this technology.

Advanced Manufacturing Technology Program

The Office of Classification and Technology Policy (OCTP) of DOE Defense Programs (DP) is initiating an aggressive technology transfer program to benefit U.S. industry. Their rationale is straightforward: Our national security depends as much on the economic strength of the United States as it does on military strength. Therefore, DOE Defense Programs are also concerned about the competitiveness of U.S. companies. DOE is willing to begin sharing its technological resources within the Nuclear Weapons Complex (NWC) with U.S. companies to the extent possible, but it also needs to acquire advanced technologies to meet its own mission requirements. Particularly in the manufacturing technology areas, U.S. companies are finding it increasingly difficult to compete with foreign companies to meet DOE procurement requirements. There is, therefore, some urgency from a DOE mission point of view to work with U.S. companies in the advanced manufacturing technology business to assist them in whatever way we can to be more competitive in the international marketplace in order to maintain both our country's economic and military strength.

DOE has funded Martin Marietta Energy Systems, Inc., to establish an Advanced Manufacturing Technology (AMT) Program, which has the mission of working with U.S. industry to transfer manufacturing technologies developed by the NWC and to collaboratively develop manufacturing technologies of mutual benefit and interest. A requisite for the success of the AMT program is close coordination and cooperation among the various components of the NWC. Since the initial focus for the AMT program will be in the area of precision flexible manufacturing systems, Energy Systems has been working with a DOE Technical Committee to help identify the technologies now available in that area for transfer to U.S. industry as well as to define collaborative projects with industry. From the viewpoint of transferring the technologies, a DOE Technology Transfer Committee will provide an advisory and communications function. Both committees have representatives from the appropriate DOE Operations Offices, Lawrence Livermore National Laboratory, Los Alamos National Laboratory, Sandia National Laboratory, Rocky Flats Plant, Kansas City Plant, and the Y–12 Plant in Oak Ridge.

The AMT program is establishing interactions with industry through several industry manufacturing technology organizations. So far, a primary organization with which the AMT program is working is the National Center for Manufacturing Sciences (NCMS), which is a not-for-profit consortium of more than 100 U.S. manufacturers and manufacturing technology suppliers. They have established a seven-member subcommittee to work with the AMT program to jointly define an R&D and technology transfer program. Currently, the following joint planning objectives are being pursued:

- Establish a strategic development plan for advanced precision flexible manufacturing systems. Determine the options for regaining the lead and develop a strategic plan to meet joint DOE and industry objectives.
- Establish a teaching factory/user facility which will provide an unclassified prototype production capability where industry personnel can learn the most advanced manufacturing technologies, new developments can be demonstrated, graduate engineering students can intern, and limited production of DOE unclassified parts can be conducted.
- Identify technologies in the precision flexible manufacturing area that are available for transfer to NCMS members. Determine how best to inform NCMS member companies of these technologies.
- Establish a manufacturing technology data base of NWC manufacturing technologies available to both DOE and industry.

The AMT program, like the partnership program with SEMATECH, is involved with industry through an existing industry R&D consortium. Rather than establish yet another consortium, we chose to establish a partnership with an existing consortium. Clearly, the AMT program is a major undertaking with many different organizations involved: the NWC, major manufacturers, manufacturing technology suppliers, and eventually universities and community colleges. Each has their own agenda. Yet each has a stake in the success of such an endeavor.

Early Observations

It is much easier to work with an existing consortium (such as SEMATECH or NCMS) than to create one of your own. We have tried to initiate two consortia ourselves so far, CAMDEC and Thermomechanical Model Software Development Center. The first failed for a complex set of reasons. The second, a twelve-company consortium, has had some turbulent times but is making excellent technological progress.

It is much easier to attract industry to a partnership if you bring both technology and money. In all the initiatives discussed above, industry dollars were substantially matched with government dollars, or other third-party dollars. The leveraged dollars assuage some of the discomfort of collaborating with your competitors. It is easier to sell an enabling process technology as a partnership technology agenda, assuming it helps all partners equally and substantially. In some cases, quantum leap improvements in the technology are required to meet strategic goals for the industry, such as the SEMATECH partnership and the Advanced Manufacturing Technology Program with NCMS. In other cases, the partnership offers a new technology opportunity, such as the High Temperature Superconductivity Pilot Center and the Optics MODIL. In all cases discussed, the focus has been on process rather than product technologies. It is much easier to sell a partnership program if the partners are seriously threatened by an outside source. If the partners perceive foreign competition, for example, as a real and substantial threat and there is some consensus about what can be done to alleviate the threat, then commitment to a partnership seems to be readily forthcoming.

Finally, the early observations offered above indicate how little we really understand about what works and what does not work in a partnership arrangement. The economic strength of our country is being severely challenged. To meet this challenge, we must learn to cooperate. Competing companies must learn to cooperate, manufacturers must learn to cooperate with their suppliers and customers, and the government must learn to cooperate with industry.

Part V

Perspectives on Japanese Consortia
and Technology Transfer

15

Outline of the Fifth-Generation Project and ICOT Activities[1]

Takashi Kurozumi

The Fifth-Generation Computer Systems (FGCS) Project was launched in 1982 as part of the information-related policy of the Ministry of International Trade and Industry (MITI). Its purpose is to research and develop a new computer technology that will provide the basis for the creation of knowledge information processing systems (KIPS) needed in the 1990s. The Institute for New Generation Computer Technology (ICOT) has been entrusted by MITI to promote this national project in cooperation with manufacturers, national and public research organizations, and universities.

The ICOT project has been proceeding according to a ten-year plan, which is divided into an initial three-year stage, an intermediate four-year stage, and a final three-year stage. The first year of the final stage was 1989. This chapter presents the outline of the FGCS project's ten-year plan, R&D results, and ICOT activities to promote the spread of R&D results.

Basic Framework of Fifth-Generation Computers

The R&D programs of the ICOT project aim at creating prototypes of fifth-generation computer systems. Conventional computers have been classified into generations according to their constituent hardware elements: vacuum tubes, transistors, IC, LSI, and VLSI. But they are all based on the same von Neumann architecture, which is characterized by sequential processing and stored-program schemes. In present-day computers, the characteristics of the architecture determine the type of machine language and software based on machine is procedural. Such computers are limited because there is an enormous gap between the way that they work and the way that human beings think, as in knowledge-based inference. Computers must follow pre-defined procedures; they cannot do processing that depends on circumstances.

Although conventional computers have architectural limitations, they have produced important technical seeds in the new, emerging technologies such as research into artificial intelligence, architecture technology, and software engineering technology. The objectives of the FGCS project are to overcome the technical restrictions of conventional computers and develop innovative computers capable of intelligent information processing. Such machines will be essential in the information-oriented society of the 1990s. There are basic needs that future computers must satisfy. They should be intelligent, easy to use, and readily available, and their software must be productive.

The concept of the fifth-generation computer systems stemmed from the idea that meeting future information processing needs would be possible by selecting existing R&D results that can benefit from further development and by combining these results in a completely new framework. The new framework (Figure 15.1) can be built by specifying a predicate logic as a new machine language, by creating a hardware system that performs highly parallel inference processing based on the new language, and by creating a software system that performs a new type of processing, a combination of the basic inference processings provided by the hardware system.

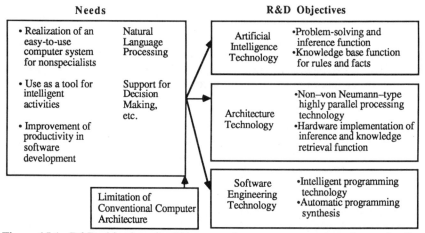

Figure 15.1. R&D objectives for fifth-generation computers to meet the needs of the coming information society of the 1990s.

Basic Structure of Fifth-Generation Computer Systems

A fundamental characteristic of intelligent activity is inference that uses every piece of stored knowledge, whether it is conscious or unconscious. Inference based on predicate logic is a procedure to extract unknown information using existing knowledge. In fifth-generation computers, hardware and software are based on a programming method called logic programming in which programs are described in the form of a logic and executed as inference. The predicate logic languages assigned to do this are called the Kernel Language (KL).

Based on the findings of previous artificial intelligence research, we estimate that fifth-generation computers will require an inference speed 1,000 times greater than conventional computers. The high-level integration provided by advanced VLSI enables us to make a reasonably compact and inexpensive computer with more than 1,000 processors working in parallel. In the ICOT project, we aim at an inference execution speed of 100M LIPS to 1G LIPS, using prototype hardware consisting of 1,000 processing elements (Figure 15.2).

1st–4th Generation	5th Generation
<von Neumann–Type>	<Non–von Neumann–Type>
• Low-level and Imperative Programming	• Logic Programming
• Numerical Processing	• Inference and Knowledge Base Processing
• Sequential Processing	• Parallel Processing

Figure 15.2. What is fifth-generation computer systems (FGCS)?

In the logic programming framework, a knowledge base for inference will also be represented in a form based on predicate logic. A relational expression in a current relational data base can correspond to a predicate logic form as its extended form. For the knowledge base function, we will start working from current relational data-base techniques and proceed to processing knowledge data that is represented in a variety of ways in the logic programming framework.

Fifth-generation computers must have a software system with basic functions needed for a knowledge information processing system which include an intelligent interaction function and an inference function that uses knowledge bases (Figure 15.3).

- The problem-solving and inference function performs meta-level inference, such as inductive inference, and is used to control hardware effectively and to solve given problems.
- The knowledge base management function acquires, stores, and uses various types of knowledge needed in the course of inference. Such a function has advanced database management capability, a knowledge acquisition capability that collects knowledge by judging whether it is meaningful, and the ability to retrieve and use knowledge effectively.

- The intelligent interface function makes computers easy to use and enables humans and computers to communicate with each other in a flexible and natural way through natural language.
- The intelligent programming function lightens the user's workload from writing programs to maintenance. It supports program development, converts given problems to more efficient programs, and verifies the accuracy of programs.

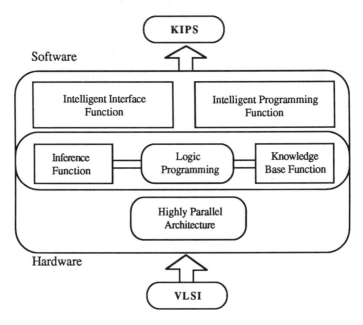

Figure 15.3. Framework of FGCS.

ICOT's aim is to realize the above four basic functions with fifth-generation computers. Although the interface between the basic software system and hardware will be implemented in Kernel Language, user languages and other languages will be defined as high-layer languages that have a modularization function and various types of knowledge representation functions.

Initiation of the FGCS Project

It was quite difficult for the FGCS Project (which was aiming at a new, creative, and risky development) to acquire a budget through government action. MITI believed that organizing a large-scale committee that included authorities from the academic community would give support for such a budget, if the committee determined that this type of computer needed to be developed. In 1979, The Fifth-

Generation Computer Research Committee, chaired by Professor Motooka, located its headquarters in the Japan Information Processing Development Center (JIPDEC). It had 35 members, including professionals from universities, national and public institutes, computer manufacturing companies, and users. A subcommittee for basic theory (organized with the ETL group as a center) brought forward the idea of an innovative computer based on a nonprocedural language and a logic programming language. A subcommittee for computer architecture advocated improving the computer along the lines of existing technology and then adding non–Von Neumann functionalities to it. The committee produced a report in August 1980 that described opposite ideas, such as improvers versus innovators and realism verses idealism.

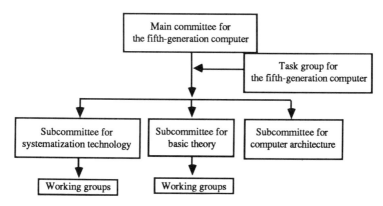

Figure 15.4. Organization of the Fifth-Generation Computer Committee.

In February 1980 the main committee held a lecture at ETL to show the possibility of non–von Neumann–type computers and to demonstrate Prolog. Its membership increased to more than 100, and working groups were organized under subcommittees on narrow technical matters. Discussions moved toward support of non–von Neumann–type computers. In its second annual report, "Proposal for the Research and Development of the Fifth-Generation Computer," the committee stated its goal by saying; "The fifth-generation computer is a knowledge information processing–oriented computer system based on an innovative methodology and techniques, able to deal with advanced functionalities, including problem solving." It described the concept as a non–von Neumann–type inference machine. The report was translated into English and distributed to computer researchers abroad.

Committee members agreed to the concept, and MITI determined to start the project as a government policy action encouraging new information processing. However, an additional process was needed to acquire a budget for the project and to formally make it a policy of Japan. The committee continued its work into 1981, and members increased to more than 120. The committee also made plans for an international conference on the fifth-generation computer to discuss the planning of the project and research results with foreign scientists, and to receive comments on the project. In October 1981 the international conference was held with more than 300 attendants, including participants from 14 countries outside of Japan. Responding to the support of MITI, the Department of Finance decided to admit the first-year budget for the project on the condition that budgets for the following years would be considered according to the results of the project's development activities.

Based on the effort and enthusiasm of a large number of people, including the members of the committee and officers at MITI, and support from many other people, research and development of the fifth-generation computer became a national project. The project has benefited from having loyal promoters, especially on the policy makers' side, since when many experts take part in project planning and compile their results in a single report, different types of requirements are evident.

Steps in the Development of the Fifth-Generation Computer

According to the report of the Fifth-Generation Computer Committee, the basic configuration of the FGCS was depicted as in Figure 15.5, and concept diagram of R&D as shown in Figure 15.6. R&D on fifth-generation computers incurs a high level of risk since it involves a large number of unknowns. Consequently, a relatively long period—ten years—has been allotted for the project. This period is divided into three stages: three years for the initial stage, four years for the intermediate stage, and three years for the final stage. The initial stage of R&D (1982–1984) was conducted with an emphasis on determining the fundamental technological elements required to build the system.

In the intermediate stage (1985–1988), the algorithms and basic architecture (to be used in subsystems that will constitute the foundations of fifth-generation systems) will be determined based on the results of the initial stage. Following this, a small-to-medium scale system will be developed using various subsystems as components.

The final stage of research and development (1989–1991), has as its goal the completion of a prototype fifth-generation system, completely integrating all the results of research performed up to this point. In addition, a primary objective of this project is the in-house development of R&D tools; this work will be carried out from the initial through the immediate project stages. Because the fifth-generation system will be based on revolutionary new programming languages, software development could not be expected to proceed efficiently using conventional computer systems. However, existing technology is being employed in the development of these high-performance tools for software development.

Organization of the Research and Development of the Fifth-Generation Computer Systems Project

In 1982, under the auspices of MITI, ICOT was founded as the central organization for promoting R&D on fifth-generation computers. The ICOT organization consists of a general affairs office and a research center. In the initial stage, the research center consisted of a research planing department and three laboratories. In the intermediate stage, it has a research planning department and five laboratories. The responsibilities of each laboratory sometimes change depending on the progress that is made.

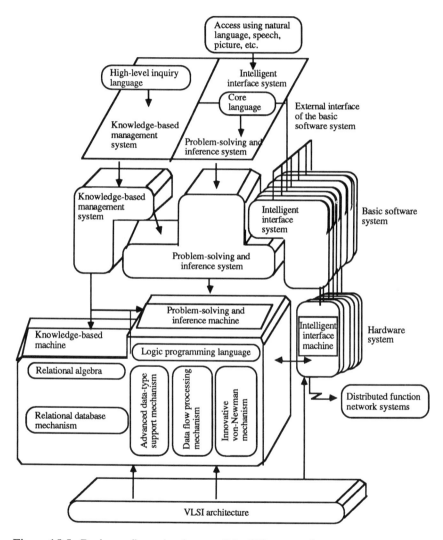

Figure 15.5. Basic configuration image of the fifth-generation computer system.

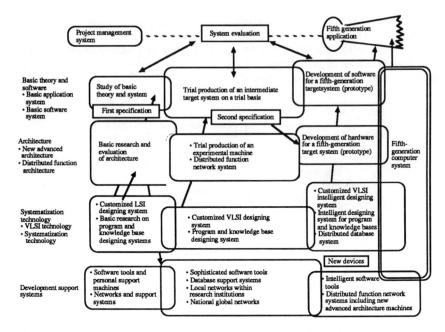

Figure 15.6. Concept diagram showing how research and development are to progress.

The research staff at ICOT is made up of researchers on loan from the Electro-Technical Laboratory, Mechanical Engineering Laboratory, NTT, KDD, computer manufacturers, and others. There were 50 staff at the beginning of the intermediate stage, and their number has increased every year. At the beginning of 1988, there were about 90 staff and soon there will be about 100. The FGCS project has been executed in an R&D organization that was formed with the idea that first-class researchers in related fields should be brought together (Figure 15.7).

MITI has set up an advisory committee to provide overall guidance concerning its research plan and R&D status of the FGCS project. Members from universities, research institutes, and companies are authorities on relevant research areas. The researchers at ICOT conduct the core R&D activities, and ICOT entrusts other R&D work, that needs experimental manufacturing and development, to manufacturers. The project is promoted as a single structure. Experts from universities and institutes have been participating in the Project Promotion Committee (PPC) and Working Groups (WGs) set up by ICOT. The PPC supplies ICOT with general advice. The WGs facilitate the exchange of information about each research subject. In 1989, there were 13 WGs that were organized according to the status of the R&D.

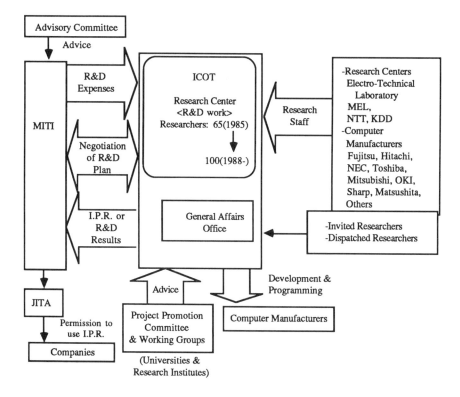

Figure 15.7. Organization of the FGCS project.

ICOT considers it important that its researchers, other researchers in Japan, and scientists from abroad stimulate each other, present their results, and exchange information. That is why ICOT actively promotes research meetings and information exchange activities with foreign research institutes such as the following:

- ICOT researchers make presentations at international conferences and publish ICOT technical reports and technical memorandums. ICOT researchers also exchange papers with foreign research institutes.
- ICOT welcomes researchers from other countries as visitors, and ICOT researchers visit research institutes and universities abroad to exchange information.
- Every year ICOT invites several renowned researchers from abroad to visit their facilities for short periods of time.
- Based on memorandums with NSF in the United States, INRIA in France, and DTI in the United Kingdom, ICOT receives researchers from these countries for extended periods of time.
- To disseminate the result of R&D activities, ICOT sponsors yearly symposiums and logic programming conferences. In addition, ICOT has

sponsored a Japan–Sweden–Italy workshop, a Japan–France AI symposium, and a Japan–United States AI symposium.

All the R&D expenses for the FGCS project are covered by the national budget. The amount is determined each year according to the government's budgeting system. The budget was 8.3 billion yen for the three-year initial stage and about 21.6 billion yen for the four-year intermediate stage (4.7 billion yen for 1985, 5.5 billion for 1986, 5.6 billion for 1987, and 5.7 billion for 1988), and 6.5 billion for 1989.

Diffusion of R&D Results

ICOT has been actively engaged in research and information exchange with many researchers and institutes that are conducting studies on subjects similar to the research themes of the FGCS project. This open policy is based on the concept that showing the results of ICOT's research to outside researchers and having free discussions with them is essential to conducting advanced research. For example, GHC (Guarded Horn Clauses), the kernel of ICOT's parallel logical languages, was created as a result of an exchange with researchers doing work on Prolog or Concurrent Prolog.

ICOT can contribute to knowledge information processing research all over the world by presenting research results and exchanging information, including preliminary ideas, with outside researchers. In addition, the presentation of research results and free research exchange is a major means to promote the diffusion of R&D results and studies on fifth-generation computer systems within and outside of Japan. Following is ICOT's plan to distribute R&D results:

- Research results are issued in materials published by ICOT such as:
  - a quarterly ICOT journal published to report the research activities at ICOT. The journal is distributed to more than 1,100 locations, 55 of which are in 36 foreign countries.
  - the results of ICOT research are compiled as technical reports (TR) or technical memos (TM) and distributed as required. There have been about 1,300 TRs or TMs, 460 of which have been written in English.
- More than 200 outside experts participate in the ICOT working groups for different research subjects. As for international research personnel, discussions with some 300 to 400 visitors are held each year. Foreign researchers are invited for short stays, and ICOT also receives researchers for long-term research programs.
- Since the entire cost of R&D activities of the FGCS project is borne by the government, intellectual property rights, including patents for the R&D results, belong to the government and are managed by AIST (Agency of Industrial Science and Technology). Any company wishing to use research results can be granted its patent by paying a fee. PSI (Personal Sequential Inference Machine) and its OS, SIMPOS, have been commercialized by companies licensed by Japan. This approach is

- expected to contribute to the establishment of a basis for the diffusion of fifth-generation computer technology.
- Software used for research which is not protected by intellectual property rights can be used at institutes outside ICOT. This process, including language specifications, is also one of the new research-promoting plans for fifth-generation computers.
- The core research activities of the FGCS project are conducted at ICOT. But the test manufacturing of hardware and software, which are highly development oriented, is entrusted to computer manufacturers. Even those developments assigned to the manufacturers are performed in a special structure, including ICOT's research because the project is designed to be integrated. Through such assignments in development activities and loans of ICOT researchers, ICOT has been giving incentives to industry to obtain an understanding of fifth-generation computer research and to prompt it to be actively involved in this research.

R&D Results of the Initial Stage (1982–1984)

R&D for the FGCS project was initially aimed at developing the basic technology required for fifth-generation computers. Within the framework of this project, R&D results concerning knowledge information processing were analyzed, and selected results were restructured to achieve the goals of the initial stage. The specific subjects of R&D included an inference subsystem, knowledge base subsystem, a basic software system, and pilot models for software development. Goals were specified independently for each project. ICOT evaluated the basic technology needed to develop fifth-generation computers by reviewing and testing various experimental systems. The major results of each R&D subject are as follows:

- *Inference subsystem:* ICOT reviewed and evaluated various inference methods, including data flow and reduction, by implementing software simulators and hardware simulators.
- *Knowledge base subsystems:* ICOT reviewed and evaluated relational databases as basic to the knowledge base function by making an experimental parallel relational database machine (Delta).
- *Basic software system:* ICOT proposed GHC as the parallel logic programming language and verified the effectiveness of the logic programming method by implementing experimental software systems, such as an experimental relational database management system (KAISER) and a discourse understanding experimental system (DUALS V.0)
- *Pilot models for software development:* ICOT showed that systems based on logic languages were viable and could be effective, by developing sequential inference machines (PSI, CHI), sequential logic programming languages (KL0, ESP), and a sequential inference machine operating and programming system (SIMPOS), although all of them were sequential.

R&D Results of the Intermediate Stage (1985–1988)

The multi-PSI system was developed as basic hardware to conduct full-scale R&D on parallel software. In the multi-PSI system, 64 processor elements are connected in a two-dimensional grid structure using special connection hardware. Following the development of this hardware, a distributed processing firmware was developed to interpret and execute the kernel language version 1 (KL1). ICOT started to use these technologies to test the manufacturing of parallel software.

Through the test manufacturing of the multi-PSI system, ICOT developed a technology to connect some 100 component processors. As the next step, aiming at PIM in the final stage, ICOT started to design hundreds of component processors and a method to connect them. At the component processor unit level, ICOT also aimed at gaining an integrity and speed four times those of the multi-PSI. The design of instruction specifications, logic design of the chip, and package designs of the board and frame were completed, and part of the test manufacturing was started.

Knowledge Base Subsystem

A test manufacturing experiment of software and hardware was conducted as a method (1) that has a data/knowledge representation suitable for logic programming, and (2) that effectively executes in parallel expanded relational operations. In research on a parallel retrieval method for knowledge bases, a parallel knowledge base retrieval mechanism test machine with eight processor elements for retrieval was created. In addition, by making a variety of test software, research was conducted on inquiry methods and management mechanisms to efficiently manage knowledge bases in a distributed environment. Through this research ICOT accumulated major component technologies required during the final stage.

Basic Software System

Concerning the kernel language version 1 (KL1) processing system, a distributed processing system according to machine language specifications was completed as firmware for the multi-PSI system. A compiler for the system description language of KL1 (KL1-C) was also completed (Figure 15.8). The first versions of PIMOS-S packaged on the pseudoparallel processing system of KL1 on PSI-II, and PIMOS-M packaged on the multi-PSI itself have also been completed. ICOT uses these technologies for test manufacturing experiments on parallel software. The basic part of a knowledge base software, preforming retrieval and management of large-scale knowledge bases, such as natural language dictionaries and expert systems, was created on PSI/PIMOS. ICOT uses it for storing natural language dictionaries. This development made a base for full-scale research on a method for knowledge description and management.

R&D on parallel languages and programming techniques (forming the core of knowledge programming) was conducted while the following basic issues were examined:

- Expansion of the functions of parallel logical languages
  - Implementation of the reflection function in GHC
  - Development of a processing system for the constrained logic programming language, CAL
- Development of a conversion method for parallel languages
  - Development of a program conversion technique for GHC
  - Development of a part calculation technique for Prolog
- Construction of theoretical models for advanced inference mechanisms, such as induction, analogy, and learning
- Enhancement of the functions of the proof support system.

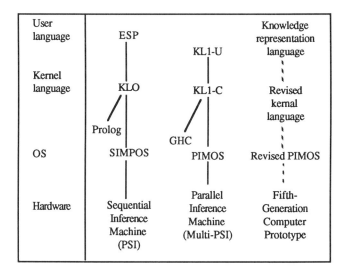

Figure 15.8. Language system of the fifth-generation computer.

There has been progress in understanding the research and development of common tools for the Japanese processing system. Version three of DUALS (Discourse Understanding Aimed at Logic-based Systems) was developed. It is a middle-sized discourse-understanding experimental system based on situation semantic theory. Also developed was an integrated environment to use a language processing system for the analysis, synthesis, and semantic description of the Japanese language, as well as to use language knowledge bases (LTB: Language Tool Box).

Outline of the Research and Development Plan for the Final Stage

The objectives of ICOT R&D in the final stage are (1) to implement prototype hardware that has a parallel architecture and that can perform high-speed inference and knowledge retrieval, and (2) to develop prototype software that can program efficiently in a parallel logic language for knowledge information processing (Figure 15.9). To achieve these objectives, ICOT will determine the organization of the project in the final stage according to the state of the R&D in the different projects. In other words, the R&D in the final stage will be geared to making a prototype system using the results of the previous research. Research will also be conducted into the basic technologies that are needed to realize the final objectives and that may be needed in the future.

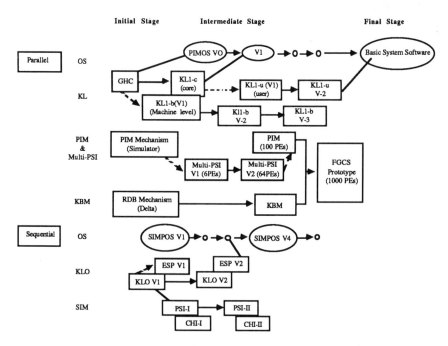

Figure 15.9. R&D steps of system software, kernel languages, and hardware.

The R&D themes in the final stage can be divided into prototype hardware and basic software. The prototype basic software system can be further divided into basic software and knowledge programming software. Prototypes of some substantiative application software systems will be developed to verify the effectiveness of various kinds of basic functions of the fifth-generation computer and to clarify the actual application.

Prototype Hardware System

The objective of the R&D for the fifth-generation computer is to implement a hardware system into which the following two mechanisms will be integrated through a hierarchical network. They are (1) dedicated hardware for realizing high-speed inference functions and (2) knowledge base functions on a vast amount of knowledge in parallel hardware architecture. The hardware system will be able to execute basic software (parallel OS) and high-speed execution of application software for large-scale knowledge information processing written in parallel logic programming language. Concretely, the hardware will be implemented by the connection of about 1,000 processing elements to provide the various functions. The hardware performance of inference operation ICOT aims for is 100M to 1G LIPS.

Prototype Basic Software System

The objectives of the R&D of the fifth-generation computer prototype basic software system are to provide the following functions:

- OS functions for an efficient parallel software execution environment by controlling and managing the hardware system.
- Functions for the description of knowledge forming the core in the development of application software in knowledge information procession.
- Functions such as cooperative problem solving and meta-level inference to support the above activities intelligently.
- Functions to construct, manage, and use knowledge bases through the structural formation of described knowledge.
- Functions for natural-language interfaces required essentially for interactive human interfaces.

This configuration consists of the prototype hardware system OS functions, basic software to provide the system programming environment, and knowledge programming software to provide the environment for natural-language processing and the description, management, and use of knowledge.

Basic Software

The basic software will handle prototype hardware control and management. Its goal is the provision of OS functions, and it consists of an inference control module for high-speed parallel inference execution control and a knowledge base management module for operation and management. The essential functions contain resource management and execution management of the prototype hardware system with an inference function and a knowledge base processing function, efficient execution management of parallel software described in parallel kernel language, and efficient operation management for storing and retrieving large-scale knowledge bases.

Knowledge Programming Software

Knowledge programming software is a group of utility software developed by using basic software. To develop application software for knowledge information processing, ICOT aims to provide a range of knowledge programming functions, a development support system, and a user interface. Research will be performed on the development of cooperative problem-solving techniques to process input problems while avoiding the conflicts and contradictions between knowledge processing software developed for different application fields. Research will also be conducted into meta-level inference functions, such as common sense decisions that approach human intelligence, and into meta-level inference techniques for the learning mechanism.

This software will provide the following functions as a step to facilitate the construction of knowledge: programming languages, various programming functions, an intelligent programming support function, a knowledge base construction function by the extraction and arrangement of expert knowledge, a function for using knowledge base efficiently according to the application, a function for reconstructing knowledge base, and all the functions required for the construction of an interactive interface that uses natural language to provide a flexible man-machine interface. This software will consist of the following three modules:

- Problem-solving and programming software module,
- Knowledge construction and utilization software module, and
- Natural-language interface software module.

Conclusion

The fifth-generation computer systems project is researching new computer hardware and software technologies using parallel architecture based on logic programming language. The goal is to produce a working prototype. ICOT is now using over 300 PSI machines and over 10 multi-PSI (64PE version) which can be divided into smaller-scale systems as R&D tools for sequential and parallel logic programming. Technical information generated by this project is being publicly released through technical reports, technical memorandums, and other publications. The project has also been the subject of international exchange. ICOT will continue such activities in the hope that it can diffuse research knowledge to everyone's benefit.

Note

1. This chapter has been written with the assistance of researchers at ICOT and with the support of MITI and many others outside ICOT proper. We wish to extend our appreciation for the direct and indirect assistance and cooperation that these entities have provided.

Appendix A. Summary of ICOT International Exchange Activities

Research Exchange Activities

United States
- Based on the agreement between ICOT and National Science Foundation (NSF), ICOT receives U.S. visiting researchers selected by NSF each year (formal agreement in June 1986).
- Dr. Tick completed a one-year posting on October 14, 1988.
- The first Japan–U.S. AI symposium was held in Tokyo from November 30 to December 2, 1987. A second symposium was held in Chicago from October 11–13, 1989, where ICOT demonstrated several software systems using multi-PSI and PSIs at the symposium.

France
- At the second Japan–France Machinery and Information Round Table (September 27, 1985), participants agreed to convene a regular Japan–France AI symposium. (The first symposium was held in Tokyo October 6–8, 1986; the second was held in Sophia-Antopolis, France, November 9–12, 1987. The third symposium was held in Izu Peninsula, Japan, November 15–18, 1989).
- Based on the agreement at the third Japan–France Machinery and Information Round Table (October 9, 1986), ICOT annually receives visiting French researchers (from six months to one year), designated by the Institut National de Recherche en Informatique et Automatique (INRIA) (formal agreement in December 1986).
- Dr. Autret and Dr. Devienne completed their postings on August 15 and December 25, 1988, respectively.
- Dr. Burg, Dr. Helft, and Dr. Dure arrived at ICOT to take up their one-year posting in January 1989.

United Kingdom
- Discussions have been in progress since 1984 regarding a research exchange between ICOT and the Alvey Directorate, which oversees the implementation of the Programme for Advanced Information Technology.

- Based on the agreement between ICOT and the Information Engineering Directorate (IED) of the department of Trade and Industry, ICOT receives U.K. visiting researchers selected by IED each year (formal agreement in December 1988).
- The first Japan–U.K. workshop was held in London, June 29–30, 1989.

Sweden
- Workshops are held with participants from ICOT and Swedish research institutes and universities, primarily the Swedish Institute of Computer Sciences (SICS). (Four workshops have been held since 1983.) This year, the Japan–Sweden–Italy workshop was held in Pisa, Italy, June 26–28, 1989.

Canada
- A research exchange agreement between ICOT and the Canadian Society for Fifth-Generation Research (CSFGR) was formalized in January 1986.

Germany
- Information exchange has been taking place since 1984 through the Japan–German Information Forum.

Invitation to Researchers Abroad

Every year, ICOT invites, at its own expense, selected researchers to Japan for one month. During this time, these scholars take part in research discussions and other related activities. To date, 51 researchers from ten countries have participated in the program.

	1982	1983	1984	1985	1986	1987	1988	1989	Total
United States	2	1	2	2	3	2	1		13
United Kingdom		4	2	2	1	1	2	1	13
France		1				1	2		4
Israel	1	1	2				1		5
Germany	1	1				1		2	5
Canada	1		1	1					3
Sweden		1				1	2		4
Italy						1	1		2
Australia				1					1
Austria					1				1
Total	5	8	8	6	5	7	9	3	51

Sponsorship of the International Conference of FGCS

FGCS'81 October 1981
FGCS'84 November 1984
FGCS'88 November 28 to December 2, 1988

ICOT Publications Distributed Abroad

ICOT Technical Reports and ICOT Technical Memoranda are regularly sent to research institutes and universities abroad, and technical information is received in exchange. These technical information exchanges take place in cooperation with 26 research institutes in the United States. The *ICOT Journal* is distributed to 550 overseas locations.

Visitors from Abroad

ICOT annually plays host to 300 to 400 visitors who are interested in learning more about the project. These include researchers, government officials, and media representatives, as well as interested people from many other fields.

16

Comparison of Several Types of Consortia for Software Research and Development

Fumihiko Kamijo

Worldwide research and development consortia have been increasing in number. There are considerable merits in R&D conducted by a group of specialists of different origins such as overcoming limitations caused by different environments in which the specialists were socialized. However, due to the short history of R&D consortia, their organizational characteristics are not well understood. A typical consortium is an organization in which the shareholders join in some R&D project. But consortia are not limited to a company-type establishment. The same type of operation can be accomplished within a division or department of the same organization. This chapter studies several Japanese R&D consortia that are operated by the same mother organization. The experience of creation, operation, and evaluation of such organizations is the focus of comparison.

Three Types of Software Research Consortia

The comparison is made among three different types of cooperative R&D establishments. The first type is a department of a nonprofit organization, an open laboratory for software technology research. Its research projects are relatively small and application based. The second type is an R&D division in a nonprofit organization that is organized for joint development of software. It integrates technology and tools for a better software development environment. The third type is a joint R&D company established by a group of private corporations for software tool development.

The Open Laboratory

The Software Technology Center (STC) of the Information-technology Promotion Agency (IPA) is the first open laboratory–type organization for software technology research in Japan. See Figure 16.1. STC was established in October, 1981, as an R&D department of IPA. It is one of the many activities based on the promotional policy of the Ministry of International Trade and Industry (MITI) as related to software industry. STC has been operated by IPA with the cooperative effort of academia, hardware manufacturers, software vendors, and computer users. The results of its R&D are open to the public with the exception of preregistered private know-how.

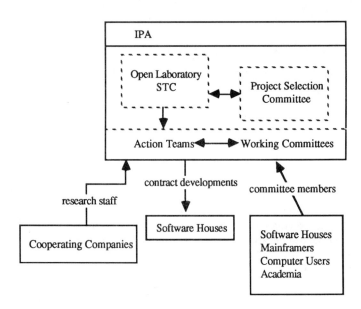

Figure 16.1. Open laboratory.

The objectives of STC are R&D in interdisciplinary subjects covering software development, software applications, and the effects of software on society. STC does not do either basic research or software product development. Rather, it tries to combine the potential of academic talent, the developmental ability of software specialists, and the application know-how of computer users. STC is an experimental department of IPA. STC's staff are either on leave from other institutions for a termed project or assigned to STC as an additional job.

The structure of STC is a collection of R&D action teams. Teams vary as to their number of technical staff and projected terms. Their average size is three to four technical staff. Their planned terms are two to three years. When a prototype program is required, the action team can ask outside organizations, for example a software house, to do the necessary development work. The average size of the total budget for an action team is the equivalent of about one million U.S. dollars including indirect costs.

A key to the success of the open laboratory system is a good selection of R&D projects. STC asks various institutions involved in information processing to propose subjects of common interest. There are no specific regulations limiting the proposing institutions. Submitted themes are then discussed by the Project Selection Committee of STC. According to the operation plan of STC, there are three major subjects of interest in selection: software development technology, software application technology, and information society–related technology.

Software Development. Subjects related to software engineering are favorite topics of academia, hardware manufacturers, and software vendors. Two major areas have been selected since 1981: requirement specification and automatic programming. Recent industrial requests to improve the quality of software resulted in the selection of the project on software evaluation. Examples of past and current STC projects are software engineering such as SKETCH (Specification, Knowledge-base, Evolution and Technology), MOTHER SYSTEM (business application generator), PAPS (Practical Automation Programming System), SNAPSHOT (visual programming based on an object-oriented paradigm), and Software Evaluation such as QFD (Quality Function Deployment, software evaluation by quality-function factor analysis) and Modified Function Point Method (Albrecht's sizing).

Software Application. Cooperative R&D projects are very appropriate for working the talents of different specialities into a team for application software. STC has completed various R&D projects in application software, such as Computer-Assisted Instruction (CAI) and Computer-Assisted Design (CAD). In CAI there is ACE (A Computerized Education, courseware executer and authoring system on micro computers) and Guidance System for Software Users (QA-type teaching material for UNIX). In CAD there is Understanding the Plant Layout Drawing (a drawing reader for plant layouts) and Understanding the Drawings of Mechanical Parts.

Information Society–Related Software. There are many problems in our computer-based society. Although STC is not a regulating agency, it influences society by developing and supplying technology to solve problems. For example, STC studies technical subjects such as computer security and natural Japanese language processing. Computer security software is exemplified by IPACS security system (access control among incompatible networks) and security software for computer viruses. Natural Japanese language processing is exemplified by IPAL (IPA Lexicon of Japanese Language for Computer), IPAL–Verbs, and IPAL–Adjectives

Management. One of the challenges of the open laboratory system is the method of management. Good R&D outcomes depend largely on the quality of leadership. Unfortunately there is no standard plan for good R&D management. Software, in particular, is a subject that is difficult to understand in terms of the level of achievement when compared with hardware-based subjects. It is essential to assign appropriate management personnel as leaders of laboratory and action teams.

Another concern of the open laboratory system is the evaluation methodology of outcomes. The marketability of the resultant technology is one measure. On this basis, STC has two subjects which have been realized as commercial products: "ACE" and "Understanding the Plant Layout Drawing." In addition, two more projects have been widely accepted by the R&D community: "IPAL" and "QFD." Both are prototypes for research use.

A third problem is staffing. Limited-term hiring is not an accepted practice in the Japanese community; consequently, "on leave" and "additional assignment" are two common mechanisms used to staff R&D consortia project teams. IPA pays direct R&D expenditures and compensates a large portion of personnel expenses, while assigning institutions have to bear the indirect and invisible losses incurred by the temporary leave of research staff. It is essential to select research subjects that are so attractive that the cooperative institutions will consider sacrificing certain research possibilities of their own. IPA provides the infrastructure necessary for the R&D environment such as lab space, computing facilities, and services to the steering and working committees.

Cooperative Division in Software Research

A good example of the division type R&D consortia is the Sigma Project, a consortium organized as an attached division to IPA under the auspices of MITI. The Sigma Project had the objective of developing a series of software tools between 1986 and 1990 with a standard interface. The tool set was then distributed commercially to help Japanese productivity in Japanese software (Figure 16.2).

Semiconductor technology has a bright outlook when it comes to the computerized society; however, a main concern is Japan's lack of ability to supply the necessary computer software. The Sigma Project has been set up to alleviate the questionable status of this advanced development environment. There are three major means of improving productivity and quality of software development: (1) integration of software tools giving a uniform platform, (2) standardization of the tools by using UNIX operating system interfaces, and (3) encouragement of program reuse through distributed processing using database and computer-networking technology

The Sigma division is planned to be an activity independent of the regular operations of IPA. The funds provided by the government and the private sector are kept in an independent account. The program products of the Sigma Project are to be propagated by an independent organization. The Sigma system is a collection of Unix workstations that are designed to meet the Sigma OS interface definition, the Sigma operating system, and Sigma software tools. Sigma tools are grouped into the following three language-oriented tool sets:

- The microprocessor-oriented system supports design, program development, program testing, and documentation phases of microprocessor based software.
- The scientific application–oriented system is a FORTRAN-based tool set which supports the full range of the software development cycle, except

for the requirement analysis phase. It has automatic code-generation tools that accept several types of chart descriptions of programs.
- The business application–oriented system is a COBOL-based tool set which supports most phases of the software development cycle. It includes a data processing pattern definition language and a COBOL program generator accepts the pattern definition.

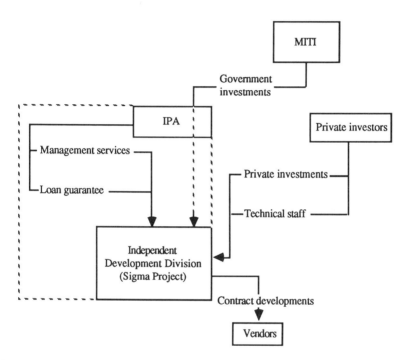

Figure 16.2. Consortium company.

The Sigma system aims at a new distributed development environment that modernizes the programmer's working environment. A programmer may continue his work at any Sigma workstation if it is connected to the Sigma network.

Management. Staffing of the division is different from that of the STC. Engineers needed for design and development of the system are assigned by member institutions, but actual development work is contracted to the selected member institutions. Staff members of the Sigma project are management oriented. By contrast, STC members have specialities and are research oriented. The funding mechanism is also different from the open laboratory type. The Sigma project is supported by government investment, private investment, and loans underwritten by IPA. Some of the loans are guaranteed by the government.

Joint Company R&D Consortia

A typical consortium for software technology development, Joint System Development Corporation (JSD), was established in 1976 as a private corporation. JSD was established by seventeen major Japanese software companies. In 1986, it was reorganized to expand its business base and the number of major shareholders was increased to nineteen (Figure 16.3).

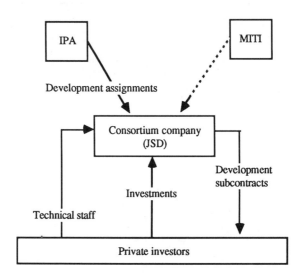

Figure 16.3. Independent division.

The JSD consortium has two major objectives: (1) development and propagation of advanced software tools for software developers and (2) cooperation in the business of large-scale software development projects. JSD is the first Japanese consortium dedicated to software development by Japanese software houses. There is no basic difference in the organization of JSD as compared to the usual commercial corporation, except for its special managing board which consists of the representatives of the nineteen major shareholders. The managing board works as the steering committee for the operating board of the company. A main difference between the independent division, such as Sigma, and a consortium is the security of development investments. In the case of the Sigma Project the mother organization underwrites necessary loans.

JSD has completed three major development efforts in software engineering technology. They are the Programming Productivity Development System (PPDS), the Software Maintenance Engineering Facility (SMEF), and the Formal Approach to Software Environment Technology (FASET).

PPDS (1976–1981) is an effort to organize various software tools based on the following system concepts: (1) integration of software tools from coding through testing, (2) supports for COBOL and FORTRAN language standards, and (3) portability of tool sets among three series of hardware (ACOS, COSMO, and M-series). The PPDS project was completed early in the era of modern software engineering. Engineers suffered from the lack of stability of new hardware, operating systems, and language compilers. Although the computing systems had much lower power than current models, a few ideas crystallized into useful tool sets. Several software tools that came out of this project are still active in the marketplace.

The PPDS project was based on the large-scale general purpose computers. Most Japanese software houses could not afford such large-scale hardware at that time. While the project was in progress, the methods and tools of software engineering had been transferred to practical use. Based on such technical advancements and on the self-evaluation of PPDS, the next project, SMEF, was designed with the following modified goals: (1) an integration of tools that are applicable to multiple phases of the software life cycle (such as design, coding, test, and maintenance), (2) application-oriented tool sets that are usable on the UNIX operating system, and (3) a workstation-based approach. SMEF products were developed according to what were very appropriate concepts in 1985. Unfortunately they have not done well in the market.

FASET (1985–1990 planned) is a project that aims at the development of upper CASE tools. It consists of the following independent subsystems: (1) ASPELA (Algebraic Specification Language); (2) FDSS (Functional Description Supporting System), a functionally descriptive language that is based on formal representation; (3) DMCASE (Design Method based on Concepts CASE tool), a system based on "Design Method Based on Concepts"; (4) STEPERS (STEPwisE Refinement Supporting tool), a pseudo-Japanese specification tool; (5) Graphtalk (Diagram-oriented design tool); (6) COORST (COmmunication ORiented Supporting Tool), a state transition model; and (7) SPECPARTNER (SPECification PARTNER), a knowledge-based specification tool.

Management. JSD has committees that control project selection and project management. The members of such committees consist of software engineers assigned to JSD by the shareholding companies. The project selection committee selects projects and prepares a proposal to MITI and IPA. Upon consensus, the project is officially recognized. It may be budgeted by MITI and designated as a project by IPA. At this stage, JSD asks the members to join the new project.

After the official assignment contract issued by IPA, the project control committee is established by JSD and the development is broken down into small subsystems. Each piece of work is reassigned to the appropriate member company. JSD also integrates the developed subsystems into a product and has the responsibility to assure a certain level of quality. In some cases, IPA has decided to go through IVV (Independent Verification and Validation). Japan's Information Processing Development Center (JIPDEC) has worked for IPA as the reviewer of IVV. The staffing of JSD has been supported by shareholding companies on a project-to-project basis, and the lack of continuity has been a problem.

Conclusion

Three different cooperative organizations for software technology research and development have been recognized by IPA. There are various pros and cons to each organizational mechanism. The following observations are tentative.

It is easier for an independent division–type organization to control a big project like Sigma. New member companies which have no prior business relationships can join in the project. A new concept for project formation may be employed, and a new style of organization using new management practices may have better results.

The open laboratory, an internal department that has a special objective, also has good characteristics. The mother organization can control the attached department much better than it can a new or more independent organization. As a consequence of close management, the internal department can obtain the well-organized services of the mother organization. Staffing is easier by using the authority of the mother organization, and direct operation allows quick decision making whenever changes are necessary.

A consortium independent of the mother organization, like JSD, enjoys freedom of management. Such an organization, made up of equal partnerships, is very powerful, but there is difficulty in that a lot of effort and time have to be consumed to obtain consensus. A consortium may expand and promote its business beyond the limited scope of the governmental assignment system. On the other hand, such expansion could be a cause of financial difficulties. There are possibilities of organizing joint ventures with shareholding companies. When shareholders are in the same type of business, it must not be forgotten that they are competitors in the marketplace. A consortium company may employ more profit-oriented management than the other two organization types

Each type of cooperative venture has advantages and disadvantages. With respect to R&D for computer software, within the limited scope of experience, the outcome of the open laboratory has given Japan some hope for good results with small projects. For a big task like the Sigma Project, an independent division, even a consortium company, seems most promising.

Note: Most of the material used to describe the characteristics of the three consortium types is based on many written reports by IPA. The operating staff of the organizations gave the author much valuable information.

17

ERATO: A Multidisciplinary Stage for Young Researchers from Industry, Government, Universities, and Abroad

Alan Engel

There are many kinds of consortia in Japan, and they defy simple description. Some people might say that Japan is one big consortium. This is, of course, not the case. However, consortia do permeate Japanese society and have for centuries.

The prevalence of consortia is evident even in traditional Japanese industries. For example, thirty years ago there were some five hundred furniture manufacturers in Tokyo. They were concentrated mainly in the Minato ward in Akasaka, Hiroo, Shiba Koen, and Azuba. Each company corresponded to one step in a modern factory and each piece of furniture was made by a consortium of storefront companies. In the 1950s and 1960s, the furniture industry was reorganized into industrial parks and large consortia. These areas are located in Tokyo near the concentrations of foreign embassies and foreign corporate employees who live in expensive apartments. Only a remnant of the old furniture industry remains.

The objectives of consortia can be divided into "hard" and "soft" objectives. Hard objectives are tangible. They include technology development and acquisition. Hard objectives also include shared facilities, as in the cases of Tsukuba Research Consortium and science parks. Soft objectives are those intangibles such as government relations, networking, researcher training, and corporate image. If there is a difference between Japanese and American views of consortia, it is in the relative weights given to hard and soft objectives. The American view puts almost all of the weight on the hard objectives. In Japan, soft objectives may outweigh the hard objectives.

Japan is inherently a networking society. The emphasis is on people. In 1988 the *Nihon Sangyo Shimbun* ran a weekly series of articles titled "Japan's Technology Cliques." It dealt only with technology leaders and their friends and acquaintances. One article traced ICOT's roots to a group of friends which formed at ETL in the late 1950s. This group included ICOT's Dr. Fuchi, Professor Aiso of Keio University, and Mr. Wada, the president of Nihon Algorithm. The series ran for fifty-three weeks and was published as a book in 1989. Japanese see technology as residing in people more than in papers or in patents.

The managing director of Keihanna Plaza, a corporation set up to service the external relations of the Kansai Science City, explained to me that Kansai Science City does not encourage Kansai industry to set up R&D facilities on location. Rather, they encourage Kansai companies to set up R&D facilities in other areas, such as Tsukuba, to improve the ties between Kansai and Tsukuba. In the U.S. setting, this would be akin to Governor Perpich of Minnesota urging 3M to place R&D laboratories in Austin because it would improve Texas-Minnesota ties. (This may have happened but I doubt it.) In Japan, good networks are valuable, long-term assets.

ERATO (Exploratory Research for Advanced Technology)

ERATO is a basic research initiative that teams researchers from industry, government, and universities to perform multidisciplinary research. ERATO is administered by the Research Development Corporation of Japan (JRDC), a statutory corporation of the Japanese government under the supervision of the Science and Technology Agency. All of ERATO's funding comes from the general national budget. U.S. companies should look closely at ERATO for opportunities in research and development and for opportunities to develop ties to the Japanese R&D community.

ERATO was established in 1981 to improve basic research in Japan by challenging several structural problems. These problems included the following:

- Widely and thinly dispersed funding of university research,
- Bureaucratic rigidity in established laboratories,
- An overemphasis on age as a basis for authority,
- Homogeneous research groups lacking the dynamic character of international laboratories,
- Traditional barriers between disciplines and between industry, universities, and government, and the
- Deemphasis of individuals.

ERATO tackled these problems by creating multidisciplinary, limited-term projects around talented scientists. An ERATO project usually includes fifteen young researchers working with an annual budget of $2 to 3 million. Project directors are selected for their ability to lead young scientists; the project director's name is attached to the project title. Young scientists are polled about whom they would most like to work for. There is at least one case in which the preferences of younger scientists resulted in a younger director being selected over an older scientist favored by ERATO's advisory board. ERATO projects are strictly limited to five years. They are carried out in rented space; JRDC has no laboratories of its own. The ERATO budget has always contained a line-item provision for foreign researchers. Intellectual property rights are shared directly with individual inventors.

ERATO places strong emphasis on unique ground-breaking research. Its research has been reviewed by the National Science Foundation's JTECH (Japanese Technology Evaluation) Program (Brinkman et al., 1988). The study found, for example, that ERATO's Hayaishi Bioinformation Transfer Project resulted in its researchers becoming the leading world experts in prostaglandin research. The Tonomura Electron Wave Project, which started October 1989, will be one of two laboratories in the world doing significant work in electron holography. (The other laboratory is at the University of Tubingen in West Germany.)

ERATO and Industry

One of ERATO's original goals was to enhance coupling between university, government, and industrial research. More than one hundred Japanese corporations have sent researchers to ERATO's projects. More than eighteen corporations have let ERATO rent laboratory space within their R&D centers. The president of ULVAC Corporation, Chikara Hayashi, personally directed one of the first ERATO projects, the Hayashi Ultra-Fine Particles Project. This year Hitachi lent ERATO one of its finest researchers, Akira Tonomura, for a project on electron holography.

Consider Nikon. Not long ago if you said "Nikon," you meant "camera." This is no longer the case. Last year Nikon's sales of semiconductor manufacturing equipment, mainly steppers, surpassed its camera sales (Figure 17.1). Nikon shipped its first stepper in 1980. Now it has 70 percent of the Japanese market. How did ERATO fit into Nikon's strategy? We can see from its research cooperation (see Table 17.1) that Nikon is clearly aiming at the state-of-the-art in precision semiconductor manufacturing equipment.

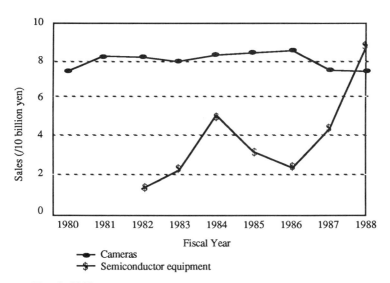

Source: Engel, 1990.

Figure 17.1. Nikon's sales of cameras and semiconductor equipment.

Table 17.1. Nikon's "Consortia" R&D

ERATO'S Yoshida Nanomechanism Project
Nanometer Displacement Equipment
X-Ray Mirror
Nanotechnology Measurements

Tsukuba University
X-Ray Microscope

Tokyo University, Tuskuba University, Kek
Scanning X-Ray Microscope

National Research Institute of Metrology (MITI)
Iodine-Stabilized Laser

Matsushita
Stepper for DRAMs

Source: Engel, 1990.

In 1985, Dr. Yoshida, the director of Nikon's precision instruments division, was chosen to be the managing director of an ERATO project in nanotechnology (Table 17.2). The project is 100 percent government funded. Fifteen researchers from ten companies and two other individuals work in three research groups. Two of these groups (fundamental analysis, and measurement and control) are located at Tsukuba Research Consortium; a third group (process) is located at Nikon's Oi Factory in Tokyo. The project has succeeded in developing measurement technology and displacement technology for one-nanometer control. It has also developed a high-reflectivity X-ray mirror. These are both key for the next generation of LSI technologies. Through ERATO, Dr. Yoshida has been able to transcend Nikon's corporate research to firmly establish himself worldwide at the forefront of nanotechnology.

Most ERATO projects are directed by university professors. One example is the Kunitake Molecular Architecture Project, which is headed by Professor Kunitake of Kyeshu University. This 100 percent government-funded project researches aggregation and self-assembly properties of amphiphile molecules. Its three research groups (structural design, functional structure, and composite structure) are located together in the Kurume Research Park in Kurume, Fukuoka Prefecture, with a satellite laboratory at the Fukuoka Prefectural Testing Laboratory in Futsukaichi. (The Kurume Research Park is the centerpiece of the Kurume-Tosu Technopolis.) This project has attracted researchers from fifteen companies, including 3M's St. Paul laboratories (Table 17.3). Of the twenty-one researchers, three are American.

Table 17.2. ERATO'S Yoshida Nanomechanism Project

Companies that sent researchers	Number of researchers
Akashi Seisakusho	2
Intel (U.S.)	1
Japan Aviation Electronics Industry	1
Kobe Steel	1
Komatsu	1
Nikon	3
Nippon Steel	1
Nitto Electric Industry	1
Seiko Electronics	1
Yasukawa Electric	3
Other Individuals	2

Source: Engel, 1990.

Foreign Researchers in ERATO

The ERATO budget has always explicitly provided for foreign researchers. For the earlier projects the budget provided for one foreign researcher per project, or about one researcher in fifteen. Project directors could increase this number and often did. More recent budgets provide for three foreign researchers per project, about twenty percent (Table 17.4). Based on recent recruiting results, there are hopes of being able to increase this percentage.

For the first seven years of ERATO, recruiting was done by the project directors through their personal contacts. This traditional method of Japanese recruiting could only bring the level of foreign researchers to about eight percent. Since 1987, ERATO management and the author (representing International Science and Technology Associates, Inc., ISTA) have been working on a second recruiting strategy through advertising in professional journals. The first advertisement appeared in January 1987 in *C&E News*. This advertisement for the Kunitake Molecular Architecture Project received about seventy applicants. The project hired the one researcher from 3M. A second round of recruiting began in September 1987 with the placement of advertisements in *Nature, Physics Today*, and *IEEE Spectrum*. Each advertisement featured one project. Four researchers were hired from thirty-five applicants.

Table 17.3. ERATO's Kunitake Molecular Architecture Project

Companies and organization with researchers in project

Asahi Chemical Industry Co., Ltd.	Asahi Glass Co., Ltd.
Kaikin Industries Ltd.	Dojindo Laboratories
Fukuoka Industrial Research Institute	Hisamitsu Pharmaceutical Co., Ltd.
Kao Corporation	Mitsubishi Chemical Industries Ltd.
Mitsui Toatsu Chemicals, Inc.	Sogo Pharmaceutical Co., Ltd.
Sumitomo Bakelite Co., Ltd.	Tokuyama Soda Co., Ltd.
Tosoh Corp.	Ube Industries Ltd.
3M (U.S.)	Mitsui Engineering and Shipbuilding Co., Ltd.

Source: Engel, 1990.

January 1989 began the third round of recruiting with a single advertisement featuring seven projects. The ad was placed in *Nature, C&E News,* and *Physics Today.* Close to three hundred worldwide researchers applied for these positions. Of the two hundred fifty-five applications received before April, about 70 percent were from the United States and Canada which emphasized that we can no longer say that American researchers are not interested in going to Japan. While only 10 percent of the applications came from Europe, European researchers account for about 40 percent of the foreign researchers in ERATO. In October 1989 ERATO let the author go to Europe for three weeks to explore better methods of recruiting. In this round, six researchers were invited to Japan for three days of interviews. Five of these researchers were hired. ISTA recruiting now accounts for ten of the twenty-eight foreign researchers in ERATO (Table 17.4).

In November 1989 we began our fourth round of recruiting with an advertisement for six projects. The ad was placed in *Nature, C&E News,* and *Physics Today.* These advertisements are being supplemented with contacts through the European organizations. The researchers recruited in this round will join ERATO in summer 1990.

ERATO is an excellent means for foreign companies and institutes to gain access to leading-edge research while developing their networks into Japanese science and technology, in addition to gaining researchers with experience in Japan. ERATO is also a way of gaining access to JRDC's follow-up development programs and consortia.

Table 17.4. Researchers in ERATO by Project

Project	Total	Foreign	Industry
Yoshida Nanomechanism (1985–90)	13	0	11
Kuroda Solid Surface (1985–90)	13	2[a]	10
Goto Quantum Magneto Flux Logi (1986–91)	15	4	7
Hotani Molecular Dynamic Assembly (1986–91)	16	3	6
Inaba Biophoton (1986–91)	15	4[a]	4
Nishizawa Terahertz (1987–92)	10	2(1[b])	5
Furusawa Morphogenes (1987–92)	11	2	3
Kunitake Molecular Architecture (1987–92)	21	3	16
Sakaki Quantum Wave (1988–93)	7	0	6
Masuhara Microphotoconversion (1988–93)	9	1(2[b])	5
Mizutani Ecochemicals (1988–93)	8	1(3[b])	3
Tonomura Electron Wavefront (1989–94)	new	new	new
Aono Atomcraft (1989–94)	new	new	new
Ikeda Genosphere (1989–94)	new	new	new
Total	138	22(6)	76
		(10 of 28 via ISTA)	

Source: Engel, 1990.
[a]Includes foreign group leader.
[b]Researchers scheduled to join ERATO.

Transfer of ERATO Research Results

The ERATO Projects have generated more than five hundred patent applications, about ninety of which are outside Japan. They have produced more than fifteen hundred papers and presentations, about four hundred of these being outside Japan. ERATO shares intellectual property rights with the inventors. The inventors on a given patent can share 50 percent of ownership if they pay 50 percent of the legal fees. Normally researchers from companies assign their rights to their company, but this is not required by ERATO. The author personally holds 50 percent rights on the U.S. patents that resulted from his work in ERATO's Ogata Fine Polymer Project.

Research results can be transferred from ERATO to industry by the four means listed in Table 17.5. By far the most common means of technology transfer is for an inventor's company to continue R&D on the technology after the inventor returns from ERATO. In two cases developments have been licensed soon after the

208 • Technology Transfer in Consortia and Strategic Alliances

project was completed. One example of licensing is the arsenic vapor pressure-controlled Czochralski method for growing GaAs crystals. This development came out of the Nishizawa Perfect Crystal Project. John Rowell of the JTECH Panel on ERATO (Brinkman et al., 1988) estimated that Mitsubishi Metals Corporation had seven GaAs crystal growth systems in place in 1988.

Table 17.5. Transfer of ERATO Research Results

Follow-up Studies at Universities and National Laboratories
(20 cases)

Example—Lattice-shape deposition of ultrafine particles
Example—Memory dynamics in asynchronous neural networks

Results Transferred to Industry

- Evaluative and developmental experiments by private companies (39 cases)
    Example—Synthesis of graphite film (Matsushita Research Institute)
    Example—Water-soluble amorphous oxides (Otsuka Chemical)
- JRDC-Industry Cooperative Development Program (2 cases)
    Example—Absorbants for separation of optical isomers (Mitsubishi Chemical)
- JRDC-Consortium R&D Program (7 cases)
    Example—Spraying process for ultrafine particles (ULVAC Corp.)
- Licensing (2 cases)
    Example—GaAs single crystals (Mitsubishi Metals)

Source: Engel, 1990.

The other two means for technology transfer to industry are through JRDC's development programs. The original purpose of JRDC at its founding in 1961 was to assist the transfer of technology from universities and government laboratories to industry. Since 1978, JRDC as handled all of the patents belonging to the national universities and government laboratories outside of MITI. (JRDC handles about a third of the MITI-AIST patents.) Part of this is done through license negotiation services. Another part is done through the JRDC-Industry Cooperative Development Program. In this program, JRDC pays for the development costs for a particular invention. If the development is successful, the contracting company pays royalties; if the development fails, there is no payment. About 36 percent (3.4 billion yen in 1987) of JRDC's total budget comes from royalties on past development projects. Table 17.5 shows examples of ERATO research results being transferred this way.

A newer program is JRDC's High-Tech Consortium Program, which was started in 1986 as a follow-up to the ERATO projects. A consortium may be based on one or more ERATO inventions in combination with inventions from other sources such as universities and government laboratories. JRDC pays a portion of the development costs, with participating companies paying the remainder. Table 17.5 shows seven cases of transfer via high-tech consortia. The High-Tech Consortium Program is still too young to evaluate or produce significant results.

Summary

There are many kinds of consortia in Japan of which ERATO is only one example. It is questionable whether lessons from ERATO can be transferred unmodified to the United States even though NSF's Emerging Technologies Initiative claims similarities to ERATO. Nonetheless, ERATO's international nature makes it an excellent entryway for American companies to Japan's R&D.

Reference

Brinkman, W., D. Oxender, R. Colwell, J. Demuth, J. Rowell, R. Skalak, and E. Wolf. "JTECH Panel Report on The Japanese Exploratory Research for Advanced Technology (ERATO) Program," NTIS, Springfield, Va., December 1988.

Part VI

Perspectives on European Consortia
and Technology Transfer

18

The European Computing-Industry Research Centre: Challenges and Answers

Hervé Gallaire

The European Computer-Industry Research Centre (ECRC) is a unique organization where major computer manufacturers cooperate in order to carry out joint research. In the present chapter, technology transfer is considered *the* measure of the success of a collaborative effort, even though there are many more reasons for pursuing collaboration between competitors.

What Is the ECRC?

The ECRC was created and is funded through the enterprise of three major European computer manufacturers: BULL, ICL, and Siemens. Its activities are intended to enhance the future competitive ability of the European information technology industry and thus to complement the work of national and international bodies.

ECRC started its operations on January 1, 1984, in Munich. It is no wonder that the Centre's field of activity was chosen to include technologies needed to improve the process of computer-assisted decision making. About six months earlier, the first major U.S. R&D consortia, the Microelectronics and Computer Technology Corporation (MCC) had begun operation in Austin, Texas. A major portion of the MCC's research program was dedicated to artificial intelligence, symbolic computing, and architectures. In Japan, ICOT was by then fully operational, and even though ICOT did not constitute a model for ECRC, its concerns are very similar. ECRC's similarity to the MCC and ICOT can be traced to similar intellectual backgrounds.

ECRC selected a technology, in a broad sense, to support its research activities. Logic programming was selected as a sophisticated representative of languages that promote a declarative style of programming. Declarativeness is a desirable feature which frees the programmer from focusing on how to solve a problem and instead lets the programmer concentrate on stating the problem to be solved. Declarative programming is seen as a way of improving productivity and thus is an important, albeit difficult, goal for the software industry. Declarative programming relies on a problem-solver system, and logic programming uses logic based techniques as its problem-solving engine. ECRC also decided to promote the concept of knowledge base as an evolution of the data-base concept by extending data bases with features usually found in logic systems. Finally, the ECRC set up the following four research groups:

- *Logic programming.* The goals of this research group are to evolve logic programming into (1) a more efficient language (work on compilation techniques, program transformation, and the like); (2) a language better suited to problem solving for real-life industrial applications; (3) a more expressive language supporting modern concepts such as those of object-oriented programming; and (4) improving the programming environment of logic, in particular through debugging techniques.
- *Knowledge bases.* The goals of the research groups are (1) to develop systems based on logic (programming) and data bases (relational) enhanced with more expressive semantic modeling features than those offered by logic and relational data bases, and (2) to devise algorithms and implement them so as to give to the knowledge base user tools for checking important properties such as integrity and consistency of the knowledge base.
- *Human-computer interaction.* In order to provide direct manipulation capabilities to the developer of a knowledge base and to its user, it was soon decided to develop tools to support the creation of these direct manipulation applications. Actually, developing such tools became a goal in itself, as they are needed for a large-class interactive applications.
- *Computer architectures for symbolic processing.* The main goal of this group is to boost the applicability of logic programming through dedicated sequential and parallel architectures.

Initially, a fifth group was set up to be dedicated to rule-based expert systems; however, it was recognized from ECRC's main studies that rule-based systems were not different enough and that ECRC results could be used in expert systems developments. Behind each of these decisions there were detailed analyses and proposals, backed up by the extensive experience of the people who defined the initial program and set of targets.

Work at ECRC is organized so as to allow projects to cross over the borders of the four areas. Basic studies are well under way in various areas that allow for early research prototype development and application simulations. ECRC was set up as a research center, working on a pre-competitive basis, aiming at developing fundamental know-how, such as theories, techniques, and support tools. It was not to be a development center since any commercial agreement was ruled out of the shareholders' decision to work together. Yet, ECRC was given as an objective to influence products of the shareholders within a five-to-ten year period. This point, however, needs to be reviewed in great depth, as it was recognized by all parties involved that this would be the most difficult challenge for ECRC.

I have very often used the following challenge during presentations of ECRC activities: Research is easy and getting interesting results is a matter of luck, but transferring results is the hardest job of all. It is a challenge because technology transfer takes dedicated teams and individuals to achieve results in very competitive fields in a short time. This is especially true when the individuals with different backgrounds and experience are recruited from a range of organization types.

The shareholder requirement to the ECRC emphasizes the obligation to prototype results in order to assess their feasibility. This requirement has in turn influenced the type of researchers recruited as ECRC made it an almost compulsory feature to have ambivalent types, interested not only in theory but also in demonstrating their own ideas. However, there have been exceptions on both ends of the spectrum. ECRC also decided not to rely on a uniform pattern for researchers. The goal was to balance experienced researchers, sometimes from remote fields of knowledge, with beginners coming out of school. The ECRC has also selected a few highly motivated staff from industry who had no research background.

ECRC, as any research center, engages in whatever theoretical and/or applied research appears necessary to solve the scientific and technical problems that arise while trying to achieve its aims. For example, the ECRC has been doing theoretical research on theorem-proving techniques for checking the consistency of sets of rules, on problem-solving techniques based on constraint propagation mechanisms, on automatic evaluation of recursive queries to a data base, on improved mechanisms for the evaluation of negation in logic programming, and on computability theory. While this research has been widely published, less theoretical research has also been conducted.

To be an industrial center, the ECRC has to evaluate the potential of its results. How feasible are the solutions? Will they lead to useless ideas or realistic systems? What benefit can the solutions bring to ECRC's shareholders? To answer such questions requires analysis, research prototype developments, experimentation, and feedback from the shareholders. An effective balance between these different types of activities is difficult to achieve. However, thanks to the temporary assignment of a significant part of its staff, the ECRC is in a good position to find and preserve such a balance.

ECRC is not devoted to specific applications development, and its work is more technology driven than application driven. On the other hand, the ECRC uses the analysis of application domains, or of specific applications, to try to comprehend the capabilities as well as the limits of the technologies it is investigating. For example, the ECRC is looking at applications in the finance and economic fields to analyze how decision-making tools could benefit from explanation techniques. Similarly ECRC has spent significant resources to investigate the domains of combinatorial optimization as well as electronic circuits design and automation tools, in order to make sure that the work done on problem solving using constraint propagation techniques is not just valid for useless applications and puzzles.

ECRC's Connections to Its Shareholders

The ECRC is an independent company registered in Munich, Germany. The consortium is managed by a shareholder's council (SC) which is made up of two members of vice-president rank from each shareholder company. The chairmanship of the SC rotates yearly on a company basis. The SC has the usual prerogatives of a board, including nominating and dismissing managers and ruling on budgets and programs. All major decisions must be unanimously approved. Indeed, each shareholder pays exactly one-third of the cost of operating ECRC.

There is a scientific advisory committee (SAC) which advises the SC on scientific and technical matters. SAC examines research proposals and results, discusses alternative proposals, and makes recommendations to approve or disapprove research programs. Research is not defined by the SAC, even though this is not formally ruled out, because it would be difficult for the three shareholder competitors to actively work on a definition of targets, and above all because the SC recognizes that research directions need to be kept constant over a certain time. The SC and SAC meet regularly, usually four times a year, which is an indication of the involvement of the companies in ECRC.

Organized Results Transfer

ECRC research is carried out on behalf of the three member companies. The rights to all research results are shared by the three companies, which have a free license on all patented results. However, since it is purely passive, this is not a sufficient mechanism to guarantee that technology transfer will take place. It assumes that "giving away" a piece of software or hardware is sufficient to transfer the knowledge embedded in it. Indeed, a more aggressive attitude toward transferring results had to be taken by the ECRC. Some suggest that a weakness in the ECRC shareholder agreement was that the word "results" had never been fully defined. On the contrary, I think that this ambiguity is a strength, not because it allows the ECRC to pretend that it transfers a lot of results (by the same token the SC could claim that there is no transfer), but because there is a rich diversity of results coming out of the ECRC, many of which cannot be easily codified.

The problem of results transfer is certainly not specific to the ECRC. However, an independent company like the ECRC has more difficulty transferring technology than does an in-house research center. There are the usual ones that result from a long-time lack of understanding between research and industry which is especially true in Europe. There are other problems specific to the ECRC situation. For instance, can the ECRC have access to marketing decisions and strategies of a particular company, or will the companies avoid giving such crucial information for fear of telling too much to the other shareholders? Some would argue that having such information is essential to guaranteeing the relevance of ECRC's research. Can ECRC establish sound relationships with the research and development units of the companies? These units can either see the ECRC as a competitor when both work on similar topics (although this could be seen as a way to maximize transfer) or ignore the ECRC if the topics selected are too different from the company's own concerns.

A number of early decisions have been made to try to ease some of the technology transfer problems at the ECRC, and, since there is not one solution, the consortium is experiencing several approaches in parallel. The first one has to do with staff transfer. There are two types of researchers at the ECRC. First, there are those assigned by the shareholders to the ECRC; these may be persons who have worked in the companies or those who were hired by the companies to be immediately assigned to the ECRC. This first type forms about 70 percent of ECRC's research staff. Their assignment is for an initial three-year period, which can be extended. Second, there are those researchers who are directly hired by the ECRC. This second type has no time limitations on their stay at the ECRC. Staff transfer is believed to be an excellent idea, but its effects have not been very spectacular. A limitation so far concerns the fact that few researchers have returned to their parent companies.

Liaison officers. Each ECRC company has assigned tasks to some of its staff, called liaison officers, to monitor each area of ECRC's research. The liaison officer is expected to identify the right partners for ECRC within their own company as well as the researchers working on related matters in case they have not been identified by the ECRCs. Such people should be interested in testing, using, discussing, and developing results from the ECRC. While this is an excellent mechanism for technology transfer, its efficiency depends on who is in charge.

Seminars. The ECRC organizes regularly scheduled technical seminars on selected topics for its shareholders. The motivation is to give a detailed presentation of the projects under way in a specific area and to have feedback from shareholder researchers, developers, and possibly staff in charge of strategy. Every company also gets to present its research, at least to some extent. These seminars provide initial personal contacts, reinforce old ties, and also provide a forum for the discussion of an entire topic. However, useful results depend on the attendance of the appropriate personnel.

Short-period exchanges. Short-period exchanges involve personnel of the shareholders spending an extended period of time at ECRC. This method of technology transfer looks very promising, and the converse is also being considered where ECRC personnel spend an extended period of time in residence at the member companies.

Joint projects. It is desirable for ECRC and shareholder companies to work together on a common set of problems and to coordinate their research. The ECRC could be used as a catalyst to drive the research, on a particular topic, of the shareholder companies. This, in fact has proved difficult to achieve. It has been working, however, in the opposite direction where the three shareholders have a joint research project (often under the auspices of a European collaborative program such as ESPRIT or EUREKA) and decide to include the ECRC. As a result, joint projects to which ECRC is invited to contribute tend to increase. This is also a measure of the increased level of collaboration of the three shareholder companies outside the ECRC. The reason why research influence works in this direction is probably due to the weight of the mechanisms put in place to influence the ECRC rather than for the ECRC to influence the research of each partner.

However, the influence of the ECRC on a partner's research does exist. Examples of such influence are numerous, where one company's research team or development team continues work started at ECRC. However, it is rare when this work is fed back to the ECRC. This one-way transfer of knowledge can be explained in two ways. One is the competition syndrome. The other is the risk of the contributing partner, which feeds back knowledge to the ECRC, seeing its research results transferred to the other companies without proper recognition or compensation. Although the shareholders' agreement in principle takes care of such cases, recent ECRC history indicates that such licensing problems pose extremely challenging problems. There is, however, a clear exception to such a situation when a project is initiated by the ECRC and has also been taken up by all three partners to pursue research and development jointly up to the production of prototypes, which are to be exploited independently.

Assessing the Transfer Process

The situation at the ECRC after five years of existence has evolved considerably and the rate of technology exchange is increasing, whether one measures it by attendance at seminars, short stays, joint projects, or informal and direct contacts. Transfers of research results from the ECRC to the shareholder companies have taken place. These transfers have been integrated into existing products and new products based on ECRC research prototypes under development.

Research results that have been transferred from the ECRC are classified according to several categories: from ideas and expertise to program code used as a basis for products and even to results which are directly marketed. For example:

- *The use of ECRC prototype code or hardware for making real products.* The ECRC CHIP problem-solving system has been turned into a product by one partner and recoded by another. The compiler SEPIA, including the graphical system KEGI and a connection to INGRES, has also been turned into a product.
- *The use of ECRC ideas embedded into prototypes for making products.* The ECRC OUTILS prototype ideas were used in PCTE, and PASTA-3 was used to define a user-interface to a data base by the same partner. EDUCE ideas were used to develop a connection between a partner's data base and its Prolog system. An ECRC compiler architecture was adopted by one partner for his own product development. And finally, the ECRC debugger was used as a basis for a partner's product.
- *The use of the ECRC prototypes to investigate new markets or gain customers.* The KCM prototype has been developed in fifty copies and will be used to test a market for high-performance systems. CHIP applications done with outside partners have been used to make product decisions and to start business relations by each partner.

- *The use of ECRC results in real applications, outside of the ECRC.*
- *The use of the image of the ECRC by all partners in dealing with their customers.*
- *The use of ECRC results at large to support or initiate projects in the companies.* While there are many such examples, few have lead to joint activities or feedback to the ECRC.

This list of ECRC research results gives an idea of the complexity of the systems that need to be put in place in order to reach a reasonable level of transfer.

Evaluation of the Process

ECRC shares the problem of how to evaluate the results of transfer with academic centers and with industrial centers as well. In Europe, academic centers have traditionally had no obligation to transfer research results, even though they are gradually taking technology transfer more seriously into their objectives. This may be particularly true on the European scene and does not apply as much to the United States and Japan. However, it is only a recent attitude in Europe and is being reinforced by the European collaborative projects.

Industrial centers have, for the most part, an obligation to transfer their results. But collaborative organizations such as the ECRC face a more difficult situation than corporate centers when it comes to technology transfer. The reason for this increased difficulty is as follows: A collaborative organization is not driven by the same forces that drive a corporate center. No matter how strong the will for collaboration, the strategic choices of large companies are not usually shared for collaboration on research. At the same time individual corporate centers are commonly informed of, and contribute to, the elaboration of strategic choices for their companies. Thus the transfer of research results of a corporate R&D center can be organized and planned ahead of time. There is a potential drawback to this: if research is fully embedded in strategy, it will leave little room for innovation.

In retrospect, I would propose some changes to the way ECRC relates to its shareholders, changes that in my view would increase, in the long term, the technology transfer benefits. Although I have focused on the role of the ECRC as a source of results for its shareholders, the research consortium has a second important objective, namely an obligation to produce scientific results. I believe that the mechanisms put in place have allowed the ECRC to work steadily toward fulfilling this objective.

Discussion

This chapter has reviewed the way a collaborative effort such as the ECRC can be organized to facilitate the transfer of results. The practical experience I have acquired in the ECRC has led to the following personal observations.

Alternatives

There is not one best solution for collaboration aimed at technology development and transfer. For example, collaboration in the European program ESPRIT does not usually involve joint laboratories. Research is done in a decentralized manner. However, there are projects where one laboratory has been set up to carry out the bulk of the program. Some other European programs, such as the electronics initiative JESSI, involve joint R&D laboratories; similarly, in EUREKA some projects have localized a significant part of their research effort.

I am convinced that a centralized effort such as the ECRC is conducive to more effective transfer of technology, in most cases, for one essential reason: The involvement of each partner in managing a central facility is much higher than in participating in a fully decentralized research and development project. Further, the financial effort required in a centralized facility that is funded directly by each partner should lead also to a more strict organization of the transfer process than in other situations. But, of course, no organizational principle is, by itself, sufficient. In order to transfer research results, one needs two partners, and no matter how willing a research team is, the end result will depend on the other partner's interest in the transfer.

Suggestions

The transfer effectiveness depends on who is involved at both ends of the process. Let us look at the obvious, natural way of organizing collaborative research efforts such as SAC for the ECRC. SAC is largely composed of people whose primary job in the companies is research. It matches the tools of the ECRC. If we look at the ECRC, its staff does not involve people with a background in industrial development or in strategy or product planning. Perhaps there should not be another body involving people who would translate strategic objectives in long-term research perspectives and who would organize, in each company, the actual transfer of results, putting in place what is needed. In the extreme, a research laboratory such as the ECRC might deal with development teams instead of the research teams of each partner and would need to structure itself so as to do this efficiently without jeopardizing its research future.

The ECRC has such an intermediate position to deal with a variety of people in its official bodies and outside of them. In particular what has been most efficient has been when one partner has high-level staff dedicated to making the transfer successful. Such a connection is more important than a joint collaborative project of ECRC and of its shareholders, financed by a third party (for example the European Commission). Let me stress this point: The single most important factor in successful technology transfer is the dedication of people involved in the process.

Another, more speculative suggestion comes from the observation that technology transfer is hindered by the lack of credibility in the technology from people who have not developed it. Although scientific collaboration between researchers is easy, transfer of results between them is difficult because of competitive feelings and the not-invented-here (NIH) syndrome. Once credibility is ensured, transfer is easier. Thus, I would like to suggest that central collaborative research centers be complemented by another structure, or by a well-identified part of the joint research facility: a development center or a development group.

I am not clear about the meaning of "development" in this sense; it could be limited to a central facility to improve the research prototypes and turn them, when needed, into development prototypes with good external visibility, such as graphics, which is so important in selling a technology. Well-polished applications could be developed and used to promote the technology. Indeed, such a facility could go one step further and become a true development shop, creating a technology customized on a contractual basis to be bought by each interested partner. This could be extended to a full-blown commercial unit, selling the technology to other customers. The idea here is to do what is done by public research centers when they spin off innovative research teams. Such a proposal might be interpreted as not fitting with having industrial shareholders whose task and goal is precisely to make the transfer happen. I have observed, however, that industry is usually more sensitive to actual results than to speculative ones.

Note: I use a different wording than "technology transfer," namely, "result transfer," because, at least in a domain like ours, technology is only one aspect of what can be transferred (if one dares to call technology what may more modestly be described as solutions and techniques). It is only an observation that there is more than technology to be transferred and to warrant such collaboration.

19

AI Research and Its Opportunities for Technology Transfer at Siemens

Dieter Schütt

High-technology companies like Siemens often have a cooperative research environment because they depend on research and innovation to remain competitive in their current markets and to expand into new markets. Such a cooperative research environment has to (1) minimize the risk potential of innovations and new "waves," (2) increase the speed with which applications of research results reach end-user markets, and (3) understand the needs of operating divisions. This includes:

- Channeling skilled people, information, and insider results of R&D into the divisions
- Ensuring against unexpected developments
- Identifying synergies
- Providing contact to the outside research community such as universities, R&D funding authorities, and other laboratories.

Cooperation at the level of European Community R&D is also important in service engineering, system engineering, pre-normative work, and experimental demonstrations in order that user needs are met by new services, which provide European industry with opportunities in new markets (see Figure 19.1).

Information technology of the nineties will require comprehensive and forward-looking research combined with an openness toward new software technologies and systems. In the next decade software will contribute considerably to the value of Siemens' products. Complex application domains have to be modeled and mastered in the context of truly large projects. Human resources, and the ability for cooperation, will be essential for a leading and not just a reactive role for Siemens in the "triad of power."

Technology transfer implies taking technology from one organization and depositing it in another. It includes the act of passing over by training, documentation, and supporting services. Transfer can mean licensing, contract research, research collaboration, joint ventures, feasibility studies, transfer projects, manufacturing, and supply agreements. The identification of recipient and collaborator relationships as early as possible is important in the transfer procedure. The technical risk potential and the complexity of a process or planned product define the role of universities, corporate R&D, and operating divisions in the transfer procedure (see Figure 19.2).

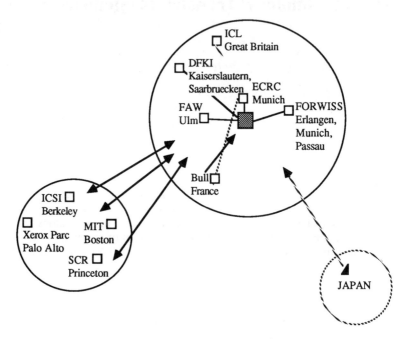

Figure 19.1. Siemens corporate R&D, information and knowledge processing: scientific connections.

Transfer by Example

Cooperation and Knowledge Transfer in the TEX-B Project

During the summer of 1985 Siemens helped establish a joint research project, TEX-B, with two research institutes and three partners from industry. In total, eighteen researchers were involved over a period of four years. The project aimed at foundations for a new generation of knowledge-based systems in scientific and technical domains. Due to the type of research and the central role of knowledge processing technology for the future, the project was funded by the German government. The first two years of the TEX-B project were used to study the work reported in the literature, prototype systems were built, and case studies for typical examples from technical domains such as electromotors were carried out. Each of the partners transfers his knowledge to a knowledge engineer, who then structures the expert's knowledge and builds the system.

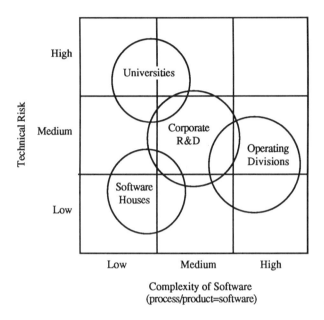

Figure 19.2. Technical risk, complexity of software, and technology transfer.

DIWA's graphical knowledge acquisition component allows the domain expert to input his structured knowledge interactively in a convenient graphical form. The knowledge has to be structured with the model of a "faulty machine." DIWA compiles these structures automatically into a knowledge base, which can be executed together with the DIWA run-time system. The model of a "faulty machine" was developed during the successful proceeding project "Night Shift Expert" and is used in DIWA for knowledge structuring. This model is a further development of a very common model in the field of heuristic diagnosis.

DIWA is applicable for building an expert system if the domain for which the expert system should be built can be structured with the model of a "faulty machine." This model is adequate for the domain "maintenance in the chip production," as was shown during the project Night Shift Expert. DIWA shortens the building process of an expert system enormously. The knowledge engineer and the doubled knowledge transformation are unnecessary. The domain expert does not need special knowledge about artificial intelligence or programming languages, due to DIWA's comfortable graphical knowledge acquisition component. Advantages of DIWA are as follows:

- The expert system can be developed and maintained by the expert.
- The development time for an expert system is shortened enormously, and the expert system can be used earlier.
- Knowledge base extension and updating is simplified.
- Knowledge base validation is easier for the domain expert.

DIWA is going to be used in different chip production sites which have 120 to 150 different types of chip production machines. Twenty to forty of these machines are crucial for productivity and are therefore intended for an expert system developed with DIWA. Knowledge bases for machines which exist at all chip production sites only have to be developed once and can be distributed among the others.

Natural-Language Access to Relational Data Bases

With the increasing complexity of today's software systems, the advent of new types of information systems, and the growing use of computers by inexperienced users, the problem of user-friendly human-machine interfaces is becoming ever more important. Besides graphical interfaces, natural language systems are an easily learnable interface, especially for those systems that demand a great expressive power such as data-base systems.

In order to bring forward advanced solutions to the problem of man-machine interfaces, Siemens is engaged in both basic research and the development of actual applications. The theoretical work encompasses such things as the development of expressive formal representations for linguistic knowledge and empirically tested, formally explicit descriptions of linguistic phenomena. The application-oriented work raises questions such as the adaptation of modern theories to a natural language data-base system.

The system SEEP (SESAM Preprocessor) is a natural language interface to relational data bases. It translates queries written in plain German into SQL statements, which are then further processed by the SESAM and INFORMIX data base handler. The current version, which contains a set of tools for adaptation of the preprocessor to a new data-base application, as well as the German SQL translator, has been used in a pilot project by an outside organization to access information on inhabitant registration in the local administration of a small town near Munich.

The SEPP interface has been developed in three phases. First, a demonstrator was developed that was tailored to the data base "SPORT," a typical customer-supplier data base on sports articles and clothes, customers and orders. The data base SPORT is the demonstration data base of the operating division for the product SESAM DRIVE, and so the link to the SESAM DBMS has been in focus.

In another step, the system has been expanded to serve both the SESAM and INFORMIX DBMS, but it is still in a demonstrator status with the DB SPORT. This configuration allowed Siemens to demonstrate a common NL-surface for a local and general DB, one on a PC under UNIX, the other under BS 2000 on a mainframe. During this phase the research department has been responsible for the pure translation from German to the different dialects of SQL, and the operating division has been responsible for the surface of the system and the micro-mainframe connection. Visitors' comments have been very fruitful for the specification of the next versions' functionality.

In the third phase, the system has been redesigned in order to make it independent of the particular application. Furthermore, a number of tools have been brought forward that allow a fast adaptation of the system. These activities resulted in a prototype that constitutes the basis for product development.

In order to test the generality of the system and the adaptation tools as well as to get more user feedback for the next version, the research department and the operating division decided to install the system for a client using either a SESAM or an INFORMIX data base. The selection criteria were (1) the customers site should be near Munich, (2) the DB should be considerably more complex than the DB SPORT, (3) the DB should be at least in first normal form, and (4) the infrastructure at the client should allow a good testing of the system. Out of four candidates, the town hall of a small city in the surrounding area of Munich has been selected. The data base has 20 relations with about 200 attributes and contains information on approximately 20,000 inhabitants. The database contains registration information like names, dates of birth and death, dates of marriages, addresses, and tax information of the inhabitants.

During the adaptation phase many minor errors in the system have been detected and corrected. It took two researchers months to carry out the pure work for adaptation. After a period of three months the installation showed that a redesign of the system's surface and parts of the NL facilities, like treatment of unknown words and error behavior, was necessary before product development could start.

The Retrieval Experiment COPSY

In the development of the Context Operator Syntax (COPSY), Siemens wanted to cope with the following problems: automatic noun-phrase recognition, automatic noun-phrase selection, automatic noun-phrase normalization (containing all variants and subphrases of a noun-phrase), noun-phrase matching, and automatic document ranking on the basis of syntactical criteria. Siemens wanted to prove the success of the approach under real-world conditions with commercial data bases.

By far the most time-consuming part of COPSY development was finding the syntactic normalization rules. They were developed on the basis of a robust morphological categorization system designed especially for information retrieval purposes for the processing of large and varying amounts of textual data. The morphological categorization system implies a word normalization component and works application-independently using small word lists (instead of complex lexica), and algorithms for suffix combinations and regularities.

COPSY was designed to work domain-independently. The COPSY retrieval was tested on several data bases: (1) retrieval of over 270,000 titles from FSTA (food science), (2) retrieval of over 200,000 index phrases from INSPEC (physics, etc.), and (3) retrieval of over 20,000 abstracts from FSTA (food science).

COPSY was part of a larger text analysis project at Siemens, called (TINA) Text-Inhalts-Analyse: text content analysis. Within TINA, a spin-off from COPSY has been tested as part of an integrated search aid and indexing system in the United States called REALIST (Retrieval Aids by Linguistics and Statistics). For this application a selection from 190,000 patent abstracts from the U.S. Patent and Trademark Office was processed by a variant of the COPSY noun and phrase normalization algorithms.

Knowledge Transfer from ECRC to Siemens

There are several possibilities for participating in the research and development results of ECRC within Siemens. Research reports are available to the people in the company, and research and development results in the form of (software and hardware) systems can be taken as a basis for refinement and extension. Here, the outcomes range from prototypes to products. In Siemens Corporate Research and Development some departments have close contact with the research groups at ECRC and are in charge of knowledge transfer. Common research within the European Community ESPRIT project between ECRC and its shareholders is a further basis for cooperation.

Two projects within Siemens Corporate Research and Development based their research on work done in ECRC: one is a Prolog extension, the other is Prolog hardware. The Prolog system, SEPIA (Standard ECRC Prolog Integrating Advanced Features), has been taken as the kernel of PROLOG-XT. Here, a modern Prolog system incorporating some useful extra features has been extended within Siemens for special application domains that are important for the company.

The envisaged domains of application are CAD electronic communication systems and XPS development. Among others, extended unification in finite algebras based on theoretical work done by Siemens could be investigated in a Prolog system and tested on real applications. A circuit verification environment was built with PROLOG-XT and installed in a Siemens ASICs design center. The results have been quite successful. The contact with the circuit designers provided input for further extensions. Some of the extensions for the prototype PROLOG-XT will become part of the Siemens Prolog product. Here, ECRC results can be found—in a modified way—in the Siemens product line.

The other example of knowledge transfer is the Knowledge Crunching Machine (KCM), a high-speed Prolog coprocessor. Based on the prototype made in ECRC, the coprocessor is now manufactured for the three shareholders in the Siemens factories. The know-how concerning this special purpose hardware is now available within the company and can be used for future systems design. A commercial department is involved in marketing and distribution activities. The machine will be available soon.

Recognition of Punched Characters on Workpieces

Industrial products often contain an inscription for identification purposes. Compared to printed characters, inscriptions from a punching machine can withstand scratches, rust, and smudges. Nevertheless punched characters are harder to read automatically. The recognition systems have to cope with various workpiece materials and illumination conditions and therefore with shadows, reflections, and noise in the pictures.

The project; "VIDEOMAT X" was started in October 1987 by the equipment production facility located in Karlsruhe. The aim was to improve the VIDEOMAT system and the industrial image processing system by making the software more intelligent. Teams were formed, one of which had to deal with the recognition of punched characters on workpieces. There have been four project partners in these teams: the energy and automation sales division for the commercial side, equipment production for the implementation of the algorithms on the VIDEOMAT hardware, a systems engineering development division for investigations of known algorithms, and innovative image processing for the research group. The new algorithms designed by the last group were tested on a general purpose computer. They were designed to be simple and work fast to fit the demands of the project partners in Karlsruhe. In spring 1989 one algorithm was chosen to be implemented on the VIDEOMAT. The tests were successful and the system may be demonstrated in the production line of a lead car-manufacturer near Munich.

Artificial Neural Nets

Recently, connectionist models have attracted considerable interest. They introduce aspects like self-organization and automatic learning of decision regions. The expectation often connected with these models is that they will be able to overcome some of the deficiencies of classical approaches. Technological advances in VLSI and computer-aided design help to emulate (massively parallel) neural networks.

In order to get reliable statements about the medium-term potential of artificial neural nets for information technology, Siemens has started the Neuro Demonstrator Project (Expenditure: approximately 60 man years). The system will combine conventional and neural approaches to industrial scene recognition. Our neural net activities are linked to the ESPRIT project PYGMALION.

At the Siemens Corp. Research Inc. (SCR) located in Princeton, an experimental acoustic motor testing system (AMTS) has been developed (see Figure 19.1). This system is used for testing automobile ventilator motors which are manufactured in the Siemens Wurzburg factory. AMTS learns from a human tester by use of examples to classify the motors into three categories (good, bad, questionable). Initial results indicate that the aim of correct automatic qualification of 80 percent will be significantly excelled. The neural network dealing with the classification has, as an input vector, a set of 32 features which represent the motor properties, three output nodes to differentiate between the categories, and 10 to 20 nodes in the hidden layer. In the learning phase the system runs parallel to a tester and registers the features as well as the human decision for each motor as a training example. For the learning phase several hundred training examples are required. For the AMTS it lasts several days. The recognition time of 0.1 to 0.2 seconds is negligibly short.

Note: The author thanks Hans-Ulrich Block, Oskar Dressler, Klaus Estenfeld, Christian Evers, Peter Muller-Stoy, Brigitte Reminger, Gabriele Schmiedel, and Bernd Schurmann for discussions on this topic.

Part VII

New Initiatives in Technology Transfer

20

The American Technology Initiative: Competitiveness through R&D Joint Ventures[1]

Syed A. Shariq

Since the beginning of the 1980s, U.S. economic competitiveness has become a slogan and symbol for many issues and concerns that continue to dominate the nation's economic policy landscape. A wide range of inquiries have produced a rich body of literature ranging from reports by the Presidential Commission on Competitiveness to the most recent *Made in America* from MIT. These publications by leading thinkers of our time cover topics such as the trade and budget deficits, education, manufacturing, quality control, cost of capital, foreign investments in the United States, international corporate alliances, technology transfer, and the like. Reading this body of literature makes one realize the challenging task that lies ahead in comprehending and effectively communicating the complex subject and in developing consensus around a set of choices for action.

The range of debate on U.S. economic competitiveness, though dealing with a complex subject matter, centers on a few common premises or themes. The task of finding our way around in this complex terrain and, more importantly, of selecting a course of action for tackling competitiveness problems can be made manageable by drawing on these common themes and developing guidelines for considering and selecting alternative choices for action. A representative set of common themes and guidelines for action is outlined in Table 20.1.

These basic themes are based on facts that the United States can no longer afford to ignore. The actions to be taken must receive serious consideration from concerned leaders dedicated to developing ways and means of enhancing U.S. competitiveness. The purpose of this chapter is to introduce one promising unilateral/independent course of action currently being explored at the American Technology Initiative (AmTech) in collaboration with the NASA-Ames Research Center. The course of action is the public/private R&D joint ventures through which we can immediately make progress in developing viable mechanisms to directly enhance U.S. economic competitiveness.

Table 20.1. Themes and Guidelines

Themes	Guidelines

• Per capita U.S. share of global wealth generated is growing at a slower rate than some of the leading industrialized competitors.
• Knowledge, capital, and infrastructure conducive to rapid change are decisive factors in gaining global competitive advantage generating wealth.
• Covergence of overall national interest with economic interest.
• Global realization of the economic benefits of the democratic free-market systems.
• Emergence and acceptance of new mechanisms (i.e., strategies, alliances, joint ventures) for the conduct of business transactions.
	• Crystallize a vision of America in the next century, well beyond the cold war, to build a shared national consciousness around a sense of mission, beliefs, values, and actions.
	• Formulate U.S. strategy to enhance U.S. economic competitiveness through:
	  - developing and implementing unilateral/independent initiatives
	  - exploring and pursuing bilateral and multilateral avenues for action.
	• Communicate a renewed sense of global reality in all of its dimensions to every U.S. citizen.
	• Impart a sense of urgency to act now.

Enhancing U.S. Competitiveness through Joint R&D Ventures

Collaborative relationships between the public and private sectors are well known and have worked well for some countries. The ability of other nations to enhance their global competitive advantage through public/private collaboration offers a suggestion to the United States. Of course, public/private collaborations in the U.S. economy are not new. There have been a range of activities in which the federal government has worked directly with the private sector as a subsidizer, insurer, underwriter, grantor, customer, or in combinations of these roles, to accomplish national objectives ranging from the development of railroads, war production, satellite systems, synthetic fuels, and to rescue major business enterprises critical to the national need. Yet, the historic reluctance on the part of most North Americans to involve government in business places mixed-sector efforts outside the U.S. norm. Traditionally, it has taken economic depressions, wartime mobilizations, or lack of private capital necessary for a principal social need to motivate direct public/private collaboration. Furthermore, across the country there is a growing concern that even though the United States spends the largest total dollar amount in the world on R&D, there is a need to further increase its direct contribution to U.S. economic competitiveness through more effective transfer and commercialization of technology.

The main advantage to be gained from public/private R&D joint ventures goes far beyond the current challenges in the transfer and commercialization of technology. Consider, for instance, the possibility of exploring, within the approximately $63 billion U.S. industrial and $65 billion U.S. government R&D in 1988, those areas where the objective of both public and private R&D efforts are almost identical as far as the specification of the technologies to be developed. By one account, published by the American Association for the Advancement of Science, this can represent as much as 10 percent of the total government-civilian R&D, or well over $2 billion in 1988.

In response, the federal government has taken steps to grant authority to federal agencies to explore the transfer and commercialization of technologies resulting from federal R&D. These steps include passing the University and Small Business Patent Procedure Act of 1980 and the Federal Technology Transfer Act of 1986, and issuing Executive Order 12591 of April 10, 1987 (Facilitating Access to Science and Technology). Further impetus has been provided through the funding and continuing support of the Federal Laboratories Consortium, the expanded role of the National Technical Information Service, establishing a manufacturing technology development focus within the National Institute for Science and Technology, and centralizing commercialization responsibility under the Commerce Department through the passage of the 1988 Omnibus Trade Bill.

These legislative actions have generated a flurry of activity within all federal agencies resulting in the formation of government, academia, and industry consortia, centers of excellence in selected high-technology areas, and new and innovative collaborative agreements between the public and private sectors. The General Accounting Office and the Department of Commerce, in May and July 1989, respectively, produced encouraging reports describing progress made by federal agencies in implementing a new form of collaborative agreement permitted under the Technology Transfer Act of 1986. These combined efforts are producing signs of hope in enhancing the transfer and commercialization of technology. However, despite these efforts, Congress and federal agencies face certain fundamental limitations in bringing these two sectors of the economy together in a market-like fashion. The existing efforts to transfer and commercialize federal technology

- are essentially adjunct or secondary to federal agency R&D missions, goals, and objectives;
- often commence after the research phase, resulting in delayed technology commercialization;
- leverage only a small fraction of the total federal R&D effort;
- rely on indirect mechanisms responding primarily to the federal technology push;
- limit the opportunity to share risks and rewards in a market-like fashion between the public and private sectors;
- do not necessarily restrict benefits to U.S. companies or universities;

- often generate mixed funding/size projects implemented on a case-by-case basis with limited uniformity of procedure or mechanisms for the private sector to efficiently and expediently interact with the government.

Now, consider the following means for overcoming these limitations: A joint venture mechanism that focuses on identifying and bringing together, in a fair market setting, those in government and industry who will cosponsor (in an equitable manner) the R&D effort at a university or a nonprofit research institution in return for receiving, respectively, technology rights for government use and exclusive rights for commercial use of the technology. Additionally, if such joint-sponsored research opportunities and resulting technologies are made available only to U.S. institutions, and if these projects commence at the inception of R&D efforts, then the United States can unearth a vast potential for public/private R&D collaboration. This will alleviate the need for Congress to appropriate additional funds or to depend on subsidization as a form of incentive. Moreover, direct federal funds or subsidies put the federal government in the difficult and undesirable position of selecting technology areas of commercial significance.

Under such a scenario, a U.S. company cosponsors (in collaboration with a federal agency) a research effort at a U.S. university or a nonprofit research institution that each would have otherwise pursued independently. The added advantages of this mechanism are

- leveraging public/private R&D resources, thus reducing the amount that each would have separately committed;
- commencing the sponsorship from the inception of R&D, thereby compressing the time to commercialization;
- increasing the probability of commercialization since, in addition to the sharing of R&D risks, the up-front industry cosponsorship represents a sound business investment;
- university and government receive appropriate royalties and have the opportunity to benefit from close collaboration with advanced industry technology and scarce talent;
- selecting the size of projects most conducive to innovation.

These advantages represent the basis of a market-like mechanism for bringing together the independent R&D interests of the public and private sectors.

Research on a prototype joint venture program along these lines has been the focus of efforts at the NASA-Ames Research Center. These efforts involved research into the legal, financial, business, and intellectual property dimensions of R&D joint ventures between government, academia, and industry. The prototype program took a "hands-on" approach with the aim of learning by doing, similar to a case study method except the R&D joint ventures implemented are R&D projects. The joint venture program experience to date confirms the basic need, strong interest, and demand for such a program.

Principles of the Public/Private Joint Venture Concept

The public/private joint R&D venture concept represents a unique effort to address current challenges to U.S. competitiveness. The joint venture participants bridge the technology gap between public and private R&D by undertaking joint ventures, sharing resources and rewards, thereby making possible those R&D projects that otherwise would not have been feasible. In this way, these joint ventures utilize market mechanisms to meet the needs of parties who directly negotiate finances, intellectual property rights, and other terms necessary to form successful joint ventures. This concept is based on the following principles:

- Joint R&D through resource and reward sharing between government, academia, and industry provides necessary and sufficient incentives and the commitment to ensure both the completion of R&D and its subsequent commercialization.
- A viable segment of mainstream public and private technology R&D has mutual and concurrent goals and objectives ideally suited for joint R&D projects.
- Technology transfer is a people-to-people process, and therefore bringing together government, academia, and industry at the inception of joint R&D endeavors strengthens the likelihood of commercialization.

In order to assure success, these joint ventures must meet certain requirements. The R&D goals of the joint venture project should be mutual and concurrent. The cost of the project must be equitably matched by the government and industry, and funded R&D should be conducted at a university or nonprofit research institution. Intellectual property rights should be divided to benefit all three participants.

The benefit to the government of joint R&D ventures is in accomplishing more with less, given the current budgetary constraints, while ensuring critically needed transfer of technology from government to academia and industry. Industry, especially small business, greatly benefits from access to university research, while reducing the cost of its product development R&D. The academic institution benefits from the opportunity to undertake research funded both by the government and industry, in addition to the commercialization potential inherent in such projects. Academia (1) will receive support for its research staff and assistance in the management of joint R&D projects, (2) will be able to publish its research, (3) will provide faculty with the opportunity to perform research leading to new products and processes, and (4) will have claim to potential royalties from commercialization of resultant technology.

Accomplishments to Date

The first joint venture project, a prototype agreement for future projects, has been negotiated and implemented. It is a three-year project to develop a miniaturized mass spectrometer. Both government and the industry participant are contributing cash and in-kind resources, and the university participant is performing the research. The cash contribution over the life of the project from all participants totals $917,000, of which the government's share is approximately $540,000. Licenses to the resulting technology will be available to the government and the industry partner. Other prototype projects explored during the last year and/or currently under exploration include:

- A joint R&D project for the development of aircraft design software. The government will contribute $450,000 over five years and software that the university will enhance. A state funding agency is contributing $162,000 over three years, and a consortium of large and small businesses will use the software under license agreements, providing approximately $500,000 over five years and feedback data to the university.
- A joint R&D project to develop a computer workstation technology for the space station. The industry participant will contribute in-kind resources and receive access to all data developed by the university. The government will contribute matching funds and will receive a prototype workstation and access to all data.
- A joint R&D project aimed at the development of a blood calcium sensor. In addition to cash, a venture capital organization offered the use of valuable patents that it has developed at a cost of over $2 million. These patents would not have been made available to the government under traditional R&D mechanisms. This three-year project anticipated $583,000 in cash contributions from the government and the industry participant, with the government's share being about $300,000. The government would have received a license to use the technology and access to all data developed during the course of the project. Although negotiated, this project will not be implemented due to the resignation of a key research scientist from the nonprofit research institution.

The estimated total government R&D contributions for these four projects is approximately $2 million, matched by $1.8 million in resource contributions from the private sector.

Over time it became obvious that in order to implement such a program on a large scale, an independent nonprofit organization dedicated to study and analyze the joint venture process and to facilitate and implement these R&D joint ventures was needed. Consequently, the American Technology Initiative (AmTech) was established to act as a focal point, model, and incubator for learning and dissemination of experience gained from the innovative R&D joint venture mechanism. AmTech's overall charter is to research those mechanisms that facilitate, negotiate, implement, and manage joint-sponsored research projects between government, academia, and industry. AmTech is established to

- conduct research on the public management of R&D;
- conduct research and explore institutional and market mechanisms for directing federal R&D, technology transfer, and commercialization efforts that will enhance U.S. industrial competitiveness through joint-sponsored research arrangements;

- promote joint-sponsored research agreements in areas of technology critical to U.S. industrial competitiveness, such as aerospace research;
- reinvest its surplus income, after allocation of funds for operating expenses, into support fellowships, R&D funding, research, and facilities to promote areas of technology critical to national needs;
- become a pioneering organization that will serve as a unique national model and will widely disseminate the results of its experience in order to support similar efforts by other federal agencies.

Joint public/private collaboration, as implemented by a nonprofit corporation, can be viewed as a U.S. model for bridging the technology gap that exists between government and industry. The sharing of risks and rewards through the joint venture mechanism can assist U.S. research, technology transfer, and commercialization efforts. The formation of AmTech as a dedicated nonprofit institution focusing on a market approach for bringing parties together represents a significant departure from many other proposals where the government is expected to play the dominant role in any public/private collaboration. This type of mechanism has more of a chance for acceptance, given the scarcity of federal funds and the historical context of U.S. institutional relationships, than those mechanisms that appear to be successful in the other countries.

Conclusion

The concept of R&D, technology transfer, and commercialization through joint public/private ventures, if implemented with wisdom and foresight, can become a driving force and catalyst for ensuring the successful commercialization of technology and can result in a unique mechanism for enhancing U.S. competitiveness in the world market. Approximately 10 percent of civilian R&D undertaken by government and industry is estimated to have both concurrent and mutual goals. This alone represents billions of dollars per year of R&D in the United States. By using a strong market-like approach to complement other efforts under way across the United States, a closer collaboration between government and the private sector can further mobilize this dormant potential in order to enhance U.S. competitiveness. The joint public/private collaboration market represents a challenge that innovative organizations like AmTech can aspire to address and undertake.

The full implementation of this program is contingent on continuing research and prototype project development experience. Issues in the areas of overall public policy, participant selection, project selection, and valuing and balancing resource contributions and technology rights among parties require further research that will result in the development of broad guidelines and procedures. Finally, this program has been successful to date because of the generous contributions of many from government, academia, and industry. It will depend on this same spirit of contribution for its future success.

Note

1. The author acknowledges the assistance of David B. Lloyd, Paul A. Masson, and Karen Risa Robbins, of American Technology Initiative, Inc., 545 Middlefield Road, Suite 170, Menlo Park, Calif. 94025, in the preparation of this chapter.

21

Technology Transfer in a Tripartite Consortium

J. Grant Brewen

The Biotechnology Research and Development Corporation (BRDC) is a consortium of seven private-sector equity members: the Agricultural Research and Development Corporation, American Cyanamid Company, Amoco Technology Company, The Dow Chemical Company, ECOGEN, Inc., Hewlett-Packard Company, and International Minerals & Chemical Corporation. BRDC is a for-profit corporation that exclusively seeks out and supports sound research projects in several sponsored institutions. It does not perform any research of its own, but functions strictly in a research management mode for the member companies.

BRDC's assets for funding research are derived from three sources: annual member company contributions, a State of Illinois research grant, and a USDA Agricultural Research Survey (ARS) grant (Figure 21.1). The total funds annually available to BRDC are four million dollars. A portion of these funds is used to defray administrative costs, and the remainder are spent to support research. The founding of BRDC was made possible by the Technology Transfer Act of 1986 and the cooperation of the ARS and the Biotechnology Center at the University of Illinois. In a sense the private sector, federal government, and the state of Illinois have entered into a cooperative research and development agreement. It is this cooperation that makes BRDC a unique consortium, in that all three entities provide funds to support fundamental research.

Research and development agreements have been entered into between the BRDC and the ARS and several public universities in the state of Illinois. A key element of these agreements is that the BRDC is assigned co-ownership of all technology developed with BRDC funds. In addition, it has the exclusive right to manage and administer the technology through licensing agreements. All revenues generated by these agreements are shared between the BRDC and the institutional entity at which the technology was developed. Although member companies have first access to technology developed through the funded research, BRDC does have the opportunity to license to third parties.

In addition to a "first look" at emerging technologies, the member companies derive a significant leveraging value to their annual contributions through the federal and state research grants. The leverage ratio is approximately twenty to one.

The BRDC membership list reveals that there is a wide diversity in markets addressed by the companies even though several of them compete in the same markets. In order to minimize competitive bidding on developed technology and to provide the maximum benefit to all members, the decision was made that the BRDC would focus its funding activities on basic research. The strategy is simply that fundamental discoveries can be used by several, or all, members for their specific future business objectives. This strategy will undoubtedly evolve into one that will contain specific focused activities in the future.

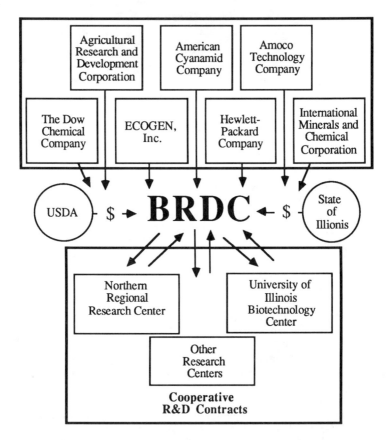

Figure 21.1. Equity members in the Biotechnology Research and Development Corporation (BRDC)

The Blending of Diverse Research Cultures

BRDC's selection of research projects to be sponsored involves the solicitation of proposals from ARS staff scientists and university faculty. These proposals are then reviewed by a Scientific Advisory Board (SAB) whose membership is comprised of two representatives from each of the member companies, two representatives from the ARS, and two faculty members from the University of Illinois. On the surface, this would appear to be an ideal method for selecting research projects to sponsor. In fact, however, the blend of individuals creates some difficulty because the cultures of the three groups are quite diverse. Academic scientists have traditionally viewed their research endeavors as their means of broadening our fundamental knowledge base, and they strive to be on the "cutting edge" of any scientific field. ARS scientists, while engaged in "basic research," are motivated, to a greater extent, by the practical applications to which their research may lead. Industrial researchers, as a result of ever-increasing pressure by top management for short-term results, are more "applied" and less prone to take the risks associated with exploratory programs.

There is no clear delineation among the three groups on the SAB. This is because many of the industrial representatives fully appreciate the need to engage in high-risk exploratory research in order to maintain their technology portfolios at the highest level of excellence, and they view BRDC as a "low-risk" opportunity to do that. There is, however, some tendency to view proposals in the light of "What will the product be and how long will it take to research the desired end point?" This attitude tends to be directly correlated to the length of time the individual has been in industrial research. The major dilemma this creates is that, when practiced to its maximum, projects are sought that have specific applications for individual member companies. This is contrary to BRDC's first principle: we want to fund research that will provide benefits to the maximum number of members.

Once the cultural and philosophical differences regarding high-risk versus practical research focus were resolved, the SAB set out to develop a research project portfolio that consisted of predominantly high-risk endeavors. As of November 1989, the BRDC is funding thirty-eight projects in six general areas. These areas can be defined as (1) process control, (2) separations, (3) biocatalysis, (4) natural products, (5) molecular biology of lower organisms, and (6) plant molecular biology. Of the thirty-eight projects, thirty are very fundamental in nature. The remaining eight have clearly defined product objectives that are attainable in three to five years.

The research is being conducted at three state universities in Illinois, three geographically separated ARS laboratories, and the National Institute of Standards and Technology. The geographical scattering of the ongoing research, coupled with the aforementioned cultural differences, presents very fundamental problems of management and communication. Being a small consortium, funding and asset-wise, BRDC must maximize the return on investment of the technology. In order to accomplish this, it was decided to attempt to manage the research projects in a quasi-industrial research management scheme, where broad objectives are identified and research teams, comprised of sponsored and member scientists, are formed that address the individual objectives. In this way it is anticipated that duplication of effort will be minimized. The minimization of duplication must include not only that between sponsored scientists, but also that between members and sponsored scientists.

Although an outline for BRDC's strategy was conceptualized early on, fleshing it out and implementing it resulted as much from serendipity as from strategic planning. The first decision was to start slowly. Consequently, having identified broad areas of interest, the BRDC funded only eleven research projects in the first year of operation. These were selected from over eighty submissions and represented what the SAB considered the best in each of the areas.

Next the BRDC implemented an "advertising campaign" to solicit proposals. This was done by having member representatives contact scientists working in the areas of interest, visiting universities and ARS laboratories, and making presentations to faculty and staff. Responsive investigators were then directed to teams of member representatives for guidance in drafting research proposals in order to have the proposals address general and specific areas of interest to the members. In addition to this method of "wiring" proposals, BRDC established a pool of funds to support highly innovative ideas that lacked, at the time of submission, substantiating data. These proposals are dubbed "exploratory research proposals" and are funded for one year to demonstrate feasibility.

In the past eight months the BRDC has funded an additional twenty-seven proposals and has two others pending the successful completion of a research agreement between BRDC and the scientists' institutions. BRDC has now reached the point where the consortium must implement the strategy for management of the research projects in a manner that will provide the greatest benefit to all members.

A critical component for the BRDC's success is the establishment of highly efficient mechanisms of information transfer to the member companies. In order to facilitate this the BRDC concluded that information transfer between sponsored scientists was critical, and that the research-team approach of management would enhance the process. The Microelectronics and Computer Technology Consortium (MCC) has designated research programs to which member companies send visiting scientists to work with MCC staff. MCC and visiting scientists are organized into research teams with specific objectives. Lacking the assets to build a research center and to staff it, the BRDC has had to develop an alternative strategy to address its primary objectives of successful technology transfer.

Research Management and Information Transfer

The central theme in any technology-based consortium is the translation of research results into technological advances and the subsequent transfer of these advances to the member companies. The development of new technology serves no useful purpose if it does not fit with the technology and business strategies of the members. In order to provide the highest probability that the research will meet those objectives several key issues must be addressed at the outset.

Although the BRDC is primarily dedicated to funding basic research, there are clear economic objectives to be achieved. To meet these objectives five central challenges have been targeted by the BRDC. The first challenge is to convince the academic scientist of the need to have clearly defined technical objectives that could be translated into a product at a future date. Stated in another way, the BRDC wants patentable results, not just publishable results. Concomitant with this, the industrial scientists have to project into the long term and not constantly be focusing on short-term results.

Another key challenge is intellectual possessiveness, i.e., the tendency toward unwillingness to share ideas and suggestions across institutional and disciplinary boundaries. A third challenge is the establishment of productive dialogue between the industrial partners and the sponsored scientists. A fourth challenge is the establishment of cooperative research activities between all three entities: academic, federal, and industrial laboratories. The fifth challenge is how to manage technology transfer most effectively once clearly defined research objectives are identified and the research has progressed to the point where it can be taken in-house by a member company.

The BRDC decided that a comprehensive and integrated approach to addressing these challenging issues was more appropriate than taking them on one at a time. To that end a plan was devised to lead to an increasingly involved dialogue between the consortium members and sponsored scientists. The components of the plan include the following:

- The first submission of proposals will be in the form of preproposals. Initial screening, based on member interest, on the basic idea occurs at this time.
- Involvement of a team of member scientists in the design and writing of the full proposal. This is to ensure that the proposal addresses the critical questions from an industrial perspective.
- Identification and assignment of key member scientists to surveillance of individual research projects. In instances where several projects are closely related the same scientist has the responsibility for all of them.
- All projects addressing the same general area of interest are grouped together for purposes of follow-up and routine progress evaluation.
- BRDC sponsorship of an annual meeting where all sponsored scientists make presentations on their work. This provides a forum for sponsoring and sponsored scientists to engage in open dialogue to exchange ideas.
- Organization of workshops at which areas of specialty are discussed in an informal atmosphere. These include sponsored scientists and their students and postdoctoral fellows who are working on the project, as well as representatives from interested member companies.
- Establishment of a seminar program in which sponsored scientists visit member companies and have the opportunity to meet with the entire technical staff.
- Initiation of a visiting scientist program. This will occur when a member company is ready to begin transferring technology in-house. The intent is to have a member scientist work in the laboratory of the sponsored scientist in order to learn the nuances of the technology and to provide additional guidance in the direction of the research.

The preproposal format has worked well. Little effort is required on the part of the author to present his/her concept in a two-page summary. If the idea appears interesting to a reasonable number of the SAB, the contributor is requested to submit a full proposal. If deemed appropriate, company representatives become involved and assist the applicant in identifying key problems to be addressed in the full proposal. In many instances the involvement of the SAB is not required, because the applicant has a clear and focused approach to addressing the research problem. However, availability of company scientists does provide the research with a pragmatic view of commercial research objectives. In those instances where proposals were modified by member suggestions, the technical objectives are more targeted with respect to potential commercial applications.

The monitoring of the research projects through annual meetings, workshops, seminars, and assignment of company liaisons has also proved valuable. There have been several instances when research objectives have been changed as a result of liaison input. In addition, the multiplicity of inputs and suggestions has avoided potential "blind alley" excursions during the course of the research.

The greatest benefits derived from the meetings and workshops have been the establishment of new collaborations between investigators and the avoidance of duplication of effort. Examples of this have been (1) the exchange of transformation vectors between molecular biologists that had been unaware of each others' work prior to a meeting and (2) the combination of experimental protocols to expedite two distinct research projects. The BRDC is currently supporting a collaborative effort involving a physical scientist and molecular biologist in developing oriented molecular arrays. This project grew out discussions at the first annual BRDC symposium, when the scientists recognized that their individual research expertise could be combined in a way that would allow intermolecular electron transfer to occur in ordered linear arrays.

In addition to these benefits, several instances of new industry/ university/federal laboratory cooperative agreements have evolved that are independent of the BRDC. These resulted from the recognition by some of the member companies that sponsored scientists were capable, and willing, to perform contact research in areas of specific interest to a company. Although none of the research has progressed to the point where member companies are ready to take it in-house, the communications network now in place should facilitate the actual transfer of the technology at the appropriate time.

The strategy, as currently visualized, for the actual process of technology transfer to occur is to establish collaborative research endeavors between sponsored scientists and member-company scientists. This would entail placing member-company scientists in laboratories at the sponsored institutions. The time frame in which this collaboration is established will be determined by the member companies. In some instances it will be judicious to do this early in the project in order to ensure that the work addresses the needs of the member companies. The ultimate success of this process will be determined by the willingness of the members to take an active role and provide the insight needed by the sponsored scientists to direct their research toward commercial applications.

Although it is too early to predict the successful transfer of emerging technology, the cooperation of the three groups of scientists in establishing open dialogue, and their willingness to enter into "team management" practices, is encouraging. This is in spite of the fact that they have highly divergent perspectives on the ultimate goals of their research.

22

A Western Technology Transfer Network for Small Business

Louis D. Higgs

The Center for the New West, which is located in Denver, is a new organization established by US WEST, Inc., in cooperation with other businesses that have headquarters or major operations in the western part of the nation. The center has been endorsed by the Western Governors' Association, the Western U.S. Senate Coalition, and the Western Interstate Region of the National Association of Counties. Its mission is to conduct research and research-driven action programs to encourage balanced growth and economic vitality in the western states, in the face of the dynamic changes of the new economy.

One major focus of the center is innovation, technology application, and commercialization. The initial emphasis has been on increasing the contribution of federal R&D efforts to business development and expansion. The center has signed memoranda of understanding with NASA and the western and midcontinent offices of the Federal Laboratory Consortium on Technology Transfer to cooperate in finding ways to make federal efforts more effective.

The center has spent over nine months examining a variety of ongoing efforts related to increasing the economic pay-off from federal R&D investments. The purpose of this examination was to try to find action items with short-term payoffs that are within the center's resources and could have a significant impact on the economy of the West. The investigation lead to two conclusions:

- First, the center should focus its efforts on the demand side, that is, on those who want to use R&D resources. There has been great progress on the supply side on the side of the federal agencies and universities that conduct R&D. This progress is expressed in significant new legislation, numbers of activities, and levels of attention by top management in federal labs and universities. While there is also increased activity on the demand side, it has not been as great or as effective.
- Second, the center needs to focus on small business and on fragmented markets because they are the most difficult areas for big R&D institutions

to respond to. The larger aggregated markets have the attention of federal labs and universities, and they have the capacity to identify and pursue opportunities in depth.

A number of experimental efforts are under way to try to address these problems. One of the more interesting of these has been developed in the western states: The Western Research Applications Center (WESRAC) at the University of Southern California, has been engaged in the development and implementation of technology transfer programs to make federal R&D information, expertise, and facilities available to the private sector. Over the past four years, WESRAC has established a state affiliates program for its NASA Industrial Applications Center Program on a cooperative basis with Business Assistance Centers located at universities in 14 western states. The purpose of this program is to make technical information, technical expertise, and other specialized technology transfer services more accessible to small businesses, and to combine such services with other business assistance services. Based on the success of this initial effort, WESRAC and its affiliates have been considering expanding the network to include nonprofit organizations involved in business assistance and economic development programs.

The experience of WESRAC and its state affiliates and the Center for the New West's recent review of technology transfer and small business assistance efforts have led to a joint venture to design, develop, and test "The Western Technology Transfer Network for Small Business," a model network to link technology transfer and small-business assistance services.

Background and Rationale for the Network

The 1980s brought two major new concepts into the arena of economic development. First is the importance of technology transfer. Second is an understanding of the major role that small entrepreneurial firms play in the growth and diversification of the U.S. economy.

Technology Transfer

Technology transfer is a relatively new term with various connotations. However, the core of the concept relates to the successful utilization of technological resources (knowledge, expertise, facilities, and actual technological developments) to strengthen the economic vitality and enhance competitiveness in a increasingly global economy. In the United States, technological resources are found in federal R&D Laboratories, universities, and in the private laboratories of businesses. Two key catalysts in both the development of this concept and the increasing realization of its economic potential were the Federal Laboratory Consortium for technology transfer (FLC), established in the early 1970s, and the NASA Office of Technology Utilization, established in the mid-1960s.

The FLC has been instrumental in the development of legislation such as the Stevenson Wydler Act and the Federal Technology Transfer Act that created a mandate and framework for federal laboratory technology transfer activities. In addition, the FLC has conducted a number of activities to strengthen the capability of federal laboratories in general and the offices of research and technology applications in the individual laboratories to provide information, advice, and assistance to businesses and public entities seeking new answers to their problems. The efforts of the FLC have been primarily directed at organizing the supply side, that is, the federal laboratories, and addressing barriers in the supply side to effective technology transfer.

Recently, the FLC has begun to focus on the demand side, that is, on the user community, and on various ways to discover and address barriers to technology transfer on the demand side. Because technology application and utilization were part of its enabling legislation, NASA has over 20 years of field experience in technology transfer. Unlike other agencies with R&D programs which had no such legislative mandate, NASA, because of its public visibility and its mandate, has been under constant pressure to develop private-sector applications and utilizations of its research and technology and to document success in their endeavors. NASA's problem was to respond to a very heavy demand, and thus time and energy went into the development of delivery system experiments and projects, such as establishing industrial applications centers in various regions of the country that provide a range of response capability to private companies and public agencies seeking to exploit NASA technology.

Small-Business Entrepreneurship

The second new emphasis in economic development is on small-business entrepreneurship. The studies of David Birch of MIT over the past 15 years have contributed to the realization of the crucial role of small business in economic development. Dr. Birch's recent work for US WEST demonstrates the degree to which small business drives the western economy. Ninety-three percent of all commercial establishments in the western states are small businesses (i.e., enterprises with 100 or fewer employees), which are generating jobs at 2.5 times the rate of large businesses, employ 52 percent of the working population in the West, and have generated 70 percent of the jobs created in the last two years.

As David Osborne and others have pointed out, states have been more responsive than the federal government to this new emphasis, not only providing extensive small-business assistance, but developing a number of new efforts such as technology innovation programs, encouraging the formation of technology incubators, and utilizing other mechanisms to strengthen innovation through small business. Traditional economic development organizations in the private sector and in state and local governments, such as chambers of commerce and local economic development corporations, are also working to develop programs focused on small business and on new technologies.

Technology Transfer and Small Business

There are two serious problems in attempting to use the highly sophisticated and organized technological resources of federal laboratories and universities to assist small businesses. First is the fragmentation of the small-business community. In the 16 western states served by WESRAC, there are over 1.7 million small businesses. Based on the experience of business assistance programs that have worked with technology transfer, it is estimated that these services are relevant to about 10 to 15 percent of small businesses. That's over 256,000 businesses in the western states. These businesses are not networked very well, much less organized, and they are relatively unaware of the available technological resources or mechanisms for accessing them.

Second is the capacity problem of the small businesses themselves. Technology transfer is not simply a technical process, it is a business process. If a technological idea, process, or product is to be transferred to a small business, that small business must have the technical, management, capital, and marketing resources to handle it. Moreover, the front-end costs of technology transfer, namely, finding and accessing relevant resources and determining whether and how to use them, are often beyond the capacity of small businesses.

The experience of WESRAC and its state affiliates and the recent state-of-the-art review of technology transfer and small business assistance efforts by the Center for the New West have convinced these organizations of the merit of conducting a major demonstration program to link technology transfer services and small-business assistance services in systematic ways. Key factors leading to this conclusion are

- Increasing emphasis on the role of small business in the economy and the need for public-sector efforts to strengthen the small-business sector and the large public investment in small business assistance programs.
- Increasing emphasis among federal agencies and universities on utilizing their expertise, facilities, existing technologies, and new technological developments for business development and expansion, the increasing public investment in technology transfer, and the difficulties encountered by federal labs in working with the small-business community.
- Evidence that linking technology transfer services and business assistance services improves the quality and effectiveness of both technology transfer and business assistance and contributes significantly to economic development at the local, state, and regional level.
- Evidence that a regional resource center for technical information and access to other technical resources provides the needed economies of scale to make the provision of technology transfer services to business assistance programs and their small business clients a sensible investment.

Technology transfer services are now available to only a very small percentage of small businesses in the West. However, there is such a large public investment in business assistance and technology transfer that a reasonably small additional investment can make a significant contribution to the benefits arising from current investments. Therefore, WESRAC and the Center for the New West have planned a demonstration program to develop and test the following model network.

The Western Technology Transfer Network for Small Business

WESRAC and the Center for the New West, in cooperation with other public and private organizations, will establish a model network: "The Western Technology Transfer Network for Small Business." This network is both modeled on and an expansion of the existing program of WESRAC and its state affiliates. Like that program, the network will not be an institution, but rather a set of cooperative relationships between existing entities. The network will consist of a set of component elements with specified responsibilities, a series of linkage mechanisms designed to strengthen the network, and a set of network maintenance and oversight mechanisms to sustain and improve performance.

Network Elements

WESRAC, utilizing its NASA Industrial Applications Center and other technical resources will function as the network's central resource for technology transfer services. The regional center provides computerized technical information services through its access to the NASA data base, other federal data bases, and over 700 public data bases. The regional center provides access to technical experts in NASA field centers and other federal laboratories through its technology counselor "tech-search" program. The center provides a brokerage function in the development of applications research agreements, joint-endeavor agreements with federal laboratories and NASA field centers and the regional center, and provides seminar and training programs in technology transfer areas.

State Associates

State associates can be any nonprofit organization providing business assistance services to small businesses within their states and/or local area. Such nonprofit organizations would include university business assistance centers, small business development centers, economic development corporations, innovation centers, incubators, chambers of commerce, agricultural extension programs, and others.

Associates will provide "work station" access to WESRAC and outreach in their local area, and will assist clients in the implementation of their projects through their normal direct services or by referral to others within the network.

Associates will also provide staff trained in the use of technology transfer programs to implement business assistance and economic development programs. In short, state associates would function like "field service agents" for technology transfer in their local area.

State Affiliate Network Service Centers

Each state shall have an affiliate network service center, typically a business assistance center which is also a state associate. These centers will serve to develop and maintain the network within the state, in addition to providing business assistance services to their own business clients. These service centers will (1) develop the intrastate network of associates; (2) provide outreach services to the associates relative to the network technology transfer programs, including development of technology transfer activity, evaluation, and benefit reporting systems for use by associates; (3) provide training programs for nonprofit associates regarding the network and technology transfer activities; and (4) develop access to other technology resources within their respective states for use by the network.

Network Linkages

Remote interactive search services will be available to all participants in the network and will enable the participant or client to work with an information expert on-line to conduct information search and retrievals. Technology counselor access will be provided via 800 phone lines for provision of "technology search" and "brokerage" services. Common training courses will be available to all participants on such topics as available information, how to use information to solve client business and technical needs, how to access technical experts, and technology transfer as an economic development tool.

A newsletter will be produced by WESRAC and circulated to all participants in the network, keeping them aware of current activities and new programs or technologies. State and regional conferences will be conducted to bring together participants for an interchange of ideas and presentation of successful activities and programs.

Network Management

Management responsibility for the network shall reside in the network operations management board. The board will be comprised of a WESRAC representative, two representatives of the network associates, two representatives of the state service centers, two business representatives who are or have been clients of the network, and a representative of the external advisory committee. The board will be responsible for network program planning and evaluation, operations oversight, fiscal control, and the preparation of reports to sponsors and the public. The board will be staffed by WESRAC and will meet twice a year. Meetings normally will be held at the site of one of the state service centers. External advice, independent assessment, and liaison with key external public and private agencies will be provided by a network external advisory committee. The committee will be comprised of sixteen members who are expected to provide several sources of strength to the network: independent knowledge and experience relative to technology transfer, small business assistance and economic development, and liaison to public and private agencies with responsibilities in those areas.

The following agencies have agreed to designate a professional to serve on the committee: the Center for the New West, the Network Operations Management Board, the Western Governors' Association, the Western Interstate Commission on

Higher Education, the Far West and Midcontinent Regions of the Federal Laboratory Consortium, NASA Office of Technology Utilization, The U.S. Department of Commerce, and the National Institute of Standards and Technology. Two private-sector sponsors of the network and two representatives of small business will also be invited to serve on the committee. The Center of the New West will staff the committee.

The External Advisory Committee will have three primary responsibilities:

- to provide guidance to the network on key issues relevant to the development and expansion of types and numbers of associates, the types of services to be provided, and the rates of expansions of both;
- to provide an independent assessment of the effectiveness and impact of the network; and
- to ensure that the programs and activities of the network support and reinforce the technology transfer and small-business assistance activities of other public and private agencies.

The committee will meet twice a year. Meetings normally will be held at the site of one of the state service centers in conjunction with the meetings of the Network Operations Management Board.

Network Scope and Funding

Based on the experience of WESRAC and its affiliates over the past four years and a review of the nonprofit organizations in the western states currently providing business assistance, it is estimated that a fully developed network would encompass over 500 state associates and provide technology transfer to over 5,000 small businesses annually.

The baseline funding for the elements of the network are currently in place and represent an annual public investment of at least $50 million. WESRAC is currently funded by NASA, DOC, USC, industry, and user fees. Nonprofit associates and state service centers are typically funded by states, SBA, EDA, DOC, respective universities, local governments, private-sector contributions, and in many cases user fees.

Based on the experience of WESRAC and its affiliates, the annual additional network operations and administrative costs are estimated at $1 million. While a major purpose of the demonstration program described below is to determine reasonable fees and additional public subsidies, it is estimated that at least 70 percent of these additional costs could be covered by a minimal fee structure.

The Demonstration

The purpose of the demonstration project is twofold: to expand the current technology transfer program of WESRAC and its state affiliates into a model regional technology transfer network which serves the entire region and all the business assistance programs in the region; and to systematically evaluate the operation of that network over a three-year period.

In order to expand the system, totally subsidized technology transfer services will be made available to small business clients of all state associate business assistance programs. Business assistance programs who wish to become state associates of the network will be required to pay a small network fee in order to ensure their commitment to utilize the network. Extensive marketing and promotion efforts will be made at two levels, to market the business assistance programs to become associates and to market the technology transfer services available through these associates to the small business community.

Evaluating the Network

Three modes will be used for evaluating the network

- *Internal evaluation:* On a continuing basis, will be based on uniform activity and result reporting. This evaluation will be a responsibility of the Network Operations Board.
- *External evaluation:* On a yearly basis, will be based on audits of the internal reports, site visits, surveys, and case studies. This evaluation will be conducted by the Center for the New West under the direction of the network external advisory committee.
- *Client evaluation:* The real test will be client evaluation in terms of (1) the decision of the associates to remain when the services are no longer fully subsidized; (2) the decision of small businesses to use the technology transfer services when they are no longer fully subsidized; and (3) the decisions of the public agencies who currently subsidize small-business assistance to continue such subsidies and include technology transfer services in such subsidies.

The three-year demonstration project will be conducted in two phases.

The Development Phase: The first year of the project will be devoted to developing the network and operating it on a minimum service level. Activities in this phase will be devoted to marketing the network, designing and conducting training programs, and focusing on the core technical services on information searches and access to federal expertise.

The Operations Phase: The second and third years of the project will be devoted to providing a full range of services, including advisory services on patents, licensing, cooperative agreements and joint ventures, expanding the expert resources available to the network to include university and industrial expertise and facilities, and systematic evaluation of the network.

Appendix I

List of U.S. Consortia Registered under the National Cooperative Research Act of 1984

Number	Venture Name/ Research Focus	Federal Registration	Number of Members
1	Exxon Production Research–Halliburton Well cementing in oil and gas wells	1-17-85	2 members
2	Software Productivity Consortium Computer software tools	1-17-85	14 members
3	Microelectronics & Computer Technology Corp. Microelectronics and computer technology	1-17-85	30 members
4	Computer-Aided Manufacturing-International Computer systems to improve production	1-24-85	95 members
5	Bell Communications Research, Inc. (Bellcore) Telecommunications technologies	1-30-85	9 members
6	Bethlehem Steel/U.S. Steel Continuous casting of steel	1-30-85	5 members
7	Semiconductor Research Corporation Semiconductors	1-30-85	41 members
8	Center for Advanced Television Studies TV transmission systems	2-1-85	9 members
9	Medium Range Truck Transmission Cooperative Project Gear transmissions	2-4-85	3 members

Number	Venture Name/Research Focus	Federal Registration Number of Members
10	Portland Cement Association Cement/concrete	2-5-85 62 members
11	Adirondack Lakes Survey Corp. Evaluation of water chemistry and fish to improve environment	2-8-85 2 members
12	Agrigenetics Agricultural genetic cell tissue research	2-8-85 1 member
13	Empire State Electric Energy Research Corp. Fossil fuel, nuclear power, electrical and environmental	2-8-85 7 members
14	Motor Vehicles Manufacturers Assn—Test Methods for Unregulated Exhaust Emissions	2-8-85 3 members
15	MVMA—Composition of Diesel Exhaust Exhaust emissions from diesel engines	2-8-85 7 members
16	MVMA—National Gasoline and Diesel Fuel Survey Variations in gasoline and diesel fuels	2-8-85 1 members
17	MVMA-Effects of Fuel and Engine Variations on Diesel Engine Emissions	2-8-85 3 members
18	MVMA/DOE Combustion Research Combustion and emission formation of intake fluid flow	2-8-85 2 members
19	MVMA—Vehicle Side Impact Test Procedure Analysis and evaluation of data re: side impacts	2-8-85 1 member
20	MVMA—CRC Motor Fuels Testing Testing to ensure compatibility between vehicles and fuels	2-8-85 3 members
21	MVMA—Fate of Polynuclear Aromatic Hydrocarbons Exhaust Dilution Sampling Systems	2-8-85 3 members
22	MVMA—Benzene Emissions Benzene fuels and emissions testing	2-8-85 3 members
23	MVMA—Fate of Diesel Particulates in the Atmosphere Atmospheric dilution of diesel particulate emissions	2-8-85 3 members
24	MVMA—Aerosol Formation in the Atmosphere Research of conversion from gaseous to particulate pollution	2-8-85 3 members
25	MVMA—Acid Rain Sources of acidity of precipitation	2-8-85 3 members
26	MVMA—Long-Range Transport of Air Pollutants Impact of transport of air pollutants on air quality	2-8-85 3 members

Number	Venture Name/Research Focus	Federal Registration	Number of Members
27	MVM-Atmospheric Transformation of Nitrogenous, Oxidant, and Organic Compounds	2-5-85	1 member
28	MVM-Truck/Trailer Brake Research Design, performance, usage and maintenance of brake system	2-8-85	15 members
29	Merrell Dow Pharmaceuticals/Hoffman-LaRoche DEMO and Interferon research on melanoma	2-19-85	2 members
30	Uninet Research and Development/Control Data Corp. Packet-switching data communications networks	3-1-85	2 members
31	Bellcore-Honeywell Inc. Gallium arsenide integrated circuits	3-25-85	2 members
32	International Fuel Cells Corp (United-Toshiba) Fuel cell systems and power plant technologies	4-5-85	2 members
33	International Partners in Glass Research Glass container development	4-10-85	7 members
34	Oncogen Limited Partnership Development of products for cancer diagnosis and treatment	4-30-85	4 members
35	Kaiser Aluminum/Reynolds Metals Co. Metallurgy and manufacturing processes for aluminum-lithium products	5-13-85	2 members
36	Plastics Recycling Foundation, Inc. Recycling of plastics	5-21-85	20 members
37	Bellcore-Avantek, Inc. High-speed integrated circuits	6-28-85	2 members
38	Bellcore-Racal Data Communications Digital connectivity for dynamic bandwidth allocation	6-28-85	2 members
39	Bellcore-U.S. Department of Army Miniature arrays of gallium arsenide crystals	6-28-85	2 members
40	Bellcore-Hertz Institut fur Nachrichtentechnik Integrated optics and optoelectronic device research	8-6-85	2 members
41	Bellcore-ADC Telecommunications, Inc. Electronically controlled integrated circuit switching matrix	9-5-85	2 members
42	Applied Information Technologies Corporation Hardware and software development for information technology	10-9-85	7 members

Number	Venture Name/Research Focus	Federal Registration Number of Members
43	NAHB Research Foundation-Smart House Development of home control and energy distribution systems	10-10-85 116 members
44	Deet Joint Research Venture DEET (pesticide) research	10-22-85 17 members
45	Geothermal Drilling Organization Technological improvements in drilling and maintaining geothermal wells	10-29-85 16 members
46	Pump Research and Development Committee Reliability and efficiency of centrifugal pumps	11-15-85 4 members
47	Battelle Memorial Institute-Optoelectronics Group Project Automated assembly of optical fibers, lasers and circuits	11-29-85 7 members
48	Bellcore-Hitachi Ltd. Optical transmission for telecommunication exchange	12-12-85 2 members
49	Intel Corp./Xicor Corp. Computer memory circuits (EEPROM)	12-12-85 2 members
50	West Virginia Univ. Industrial Cooperative Research Center Fluidization and fluid particle science	12-17-85 5 members
51	Subsea Production Maintenance Joint Industry Program Downhole maintenance on subsea wells	1-14-86 12 members
52	CharTech: Kean Manufacturing and Fabristeel Products Self-piercing metal fasteners	1-28-86 2 members
53	Norton/TRW Ceramics Ceramic products, composites and coatings	1-28-86 2 members
54	Southwest Research Institute Heavy-duty diesel particulate trap regeneration	2-18-86 29 members
55	Petroleum Environmental Research Forum Pollution control and waste treatment technology for petroleum	3-14-86 18 members
56	Pyrethrin Joint Research Venture Toxicological research on Pyrethrin (pesticide ingred.)	3-18-86 9 members
57	Keramont Research Corporation Applications of new technology for advanced ceramics	4-3-86 6 members

Appendix I • 259

Number	Venture Name/ Research Focus	Federal Registration Number of Members
58	Corporation for Open Systems International Products and services for open network architecture	6-11-86 41 members
59	Southwest Research Institute: Remaining Life Methodology for Disc Rim Cracking Project	6-11-86 4 members
60	Joint Venture: Continuous Casting of Steel Sheet Continuous casting of steel sheets	6-12-86 4 members
61	International Magnesium Development Corporation Magnesium technology	6-30-86 12 members
62	Wickes Manufacturing Company/Cycles Peugeot Seating systems and components for automobiles	7-15-86 2 members
63	Engine Manufacturers Association Diesel emission control systems	7-17-86 18 members
64	ARCO Chemical Co./Air Products & Chemicals Polyalkylene carbonate compositions	8-28-86 2 members
65	Industry/University Center for Glass Research Glass manufacturing and melting	9-10-86 12 members
66	ADBAC QUAT Joint Venture Pesticide ingredients	10-7-86 8 members
67	Babcock & Wilson Co./Seitz-Filter-Werke Filtration processes and devices for power plant application	12-24-86 2 members
68	Pine Oil Joint Research—Am. Cyanamid Co. Pine oil in pesticides	2-5-87 10 members
69	Industry/University Cooperative Research Center for Software Engineering Improve productivity of computer software developers and quality of software	2-9-87 10 members
70	Bellcore/Fujitsu Network architecture for telecommunications systems	2-13-87 8 members
71	Bellcore: IN/2 Venture Network architecture for telecommunications systems	2-13-87 8 members
72	National Center for Manufacturing Sciences Improvement of manufacturing processes and materials	3-17-87 106 members
73	Petroleum Environment Research Forum 86–06 "Evaluation of Hazardous Waste Solidification Process" stabilization of refinery hazardous wastes	3-25-87 6 members
74	Petroleum Environment Research Forum 86–09 "Microbiological Processing of petroleum Oily Wastes" Microbiotic degrad. of petroleum oily wastes	3-25-87 7 members

Number	Venture Name/ Research Focus	Federal Registration Number of Members
75	Petroleum Environment Research Forum 86–05 Biogradative treatment of spilled petro. products	3-25-87 2 members
76	Metal Casting Technology, Inc. Metal casting technologies, processes, materials, and equipment	4-1-87 2 members
77	Ziroconium Alloy Tubing-Sandvik Special Metals Corporation Corrosion of zirconium alloy tubing	4-24-87 5 members
78	Bellcore–TroQiomt Exchange and exchange access of GaAs integrated circuits	4-3-87 2 members
79	CPW Technology TV and motion picture technologies and copyrights	6-15-87 3 members
80	Pacific Bell/Integrated Network Corporation Exchange, exchange access and data telecommunications	7-1-87 2 members
81	Bellcore/Microwave Semiconductor Corporation Compound semiconductor materials	7-13-87 2 members
82	Corning Glass Works/Nippon Telegraph and Telephone Fluoride glasses used in optical waveguide fibers	7-15-87 2 members
83	MVMA—Fluorocarbon-134a Lubricants for mobile A/C systems	7-30-87 2 members
84	MVMA—Hose Connections Hose materials in mobile air conditioning systems	6-30-87 2 members
85	Southwest Research Institute—Lubricating Oil Project fuel composition and quality for year 2000	8-26-87 6 members
86	Material Handling Research Center Materials handling	9-11-87 33 members
87	Southwest Research Institute—Lubricating Oil "Importance of Lubricating Oil in Diesel Particulate Emissions"	9-18-87 12 members
88	Bellcore-Vitesse Semiconductor Corporation Gallium arsenide substrates in integrated circuits	10-2-87 2 members
89	Joint Venture of All-Terrain Vehicle Dist. Development of voluntary safety standards	10-14-87 5 members
90	National Forest Products Association Reliability-based design manual for wood	10-30-87 11 members
91	Berkeley Sensor and Actuator Center Sensors and actuators	12-15-87 9 members

Number	Venture Name/ Research Focus	Federal Registration Number of Members
92	Bellcore-NEC Optical devices	12-18-87 2 members
93	Composite Materials Characterization, Inc. Testing and grading of composite materials	1-15-88 7 members
94	West Agro-Iodophors Joint Venture Toxicological research on antimicrobials	2-19-88 15 members
95	Bellcore-NEC Broadband technology and SONET format	2-19-88 2 members
96	Bellcore-Sumitomo Electric Industries Semiconductor materials and devices for electronics	4-6-88 2 members
97	Biotechnology Research and Development Corporation Biotechnology	5-12-88 7 members
98	SEMATECH Advanced semiconductor manufacturing techniques	5-19-88 14 members
99	Industry/University Cooperative Research Center for Microwave/Millimeter-Wave Computer-Aided Design Integrated system design for microwave methods and tools	5-31-88 11 members
100	Bellcore-Nippon Hoso Kyokai HDTV R&D	6-3-88 2 members
101	Bellcore-David Sanoff Research Center Superconducting materials, high-speed systems	6-3-88 2 members
102	Petroleum Environment Research Forum, No. 88–01 "Premix Surface Combustion Burner Demo. Program" Retrofit burners in refinery heaters	6-3-88 19 members
103	Institute for Manufacturing and Automation Research Automation and robotics	6-30-88 8 members
104	International Diatomite Producers Association Diatomite information	7-14-88 6 members
105	Manville Corp.-Bird, Inc. Roofing Div. Agreement Asphalt shingle recycling	7-18-88 2 members
106	Microelectronics Center of North Carolina Microelectronics	8-1-88 17 members
107	Automotive Polymer-Based Composites Partnership Composites for vehicles and components	8-4-88 3 members
108	National Forest Products Association Design specifications for wood	8-4-88 16 members

Number	Venture Name/ Research Focus	Federal Registration Number of Members
109	Fabric Softener Quats Joint Venture Collection of data on ammonium compounds	8-19-88 4 members
110	Dialkyl Project Toxicological research on ammonium compounds	8-25-88 4 members
111	Industry/University Cooperative Research Center for Simulation and Design Optimization of Mechanical Systems	8-31-88 13 members
112	Open Software Foundation, Inc. Open environment for software applications	9-7-88 180 members
113	Cable Television Laboratories, Inc. Technical information and development of technology for cable TV	9-7-88 21 members
114	Bellcore-Landis and Gyr Engineering problems of network interfaces	9-15-88 2 members
115	Bellcore-Telettra Video transmissions	9-15-88 2 members
116	Southwest Research Institute: High Temperature Diesel Particulate traps	9-27-88 13 members
117	PDES, Inc. Development of standard product data exchange specifications	10-14-88 21 members
118	Southwest Research Institute: Bus Emissions Tech. Cooperative Industry Project Environment-acceptable bus systems	10-21-88 4 members
119	Measurement and Control Engineering Center Measurement and control research	11-4-88 7 members
120	Bellcore-Graphics Communications Technologies Video technology exchange and access services	11-16-88 2 members
121	Bellcore-Fujitsu Telecommunications services and systems	11-16-88 2 members
122	X-Open Ltd. Multi-vendor computer systems and programs	11-16-88 20 members
123	Industrial Consortium for Research and Education Material and energy flow and hydrogeology	12-8-88 4 members
124	OSI/Network Management Forum Reference model for open systems networks	12-8-88 95 members
125	Petroleum Environment Research Forum, No. 87–05 Microbiological degrad. of petroleum oil sludges	12-30-88 5 members

Number	Venture Name/ Research Focus	Federal Registration Number of Members
126	Omega Marine Services International, Inc. Deep-water mooring technology	3-1-89 8 members
127	UNIX International, Inc. Unix operating systems	3-1-89 122 members
128	B. F. Goodrich-European Vinyls Corporation Vinyl chloride and PVC resins	3-10-89 2 members
129	CAD Framework Initiative, Inc. Design automation frameworks	3-13-89 98 members
130	Southwest Research Institute: Wet Welding at Greater Depth Wet welding processes	3-13-89 9 members
131	Southwest Research Institute: Investigation of Effects of Mechanical Aids on the Annular Flow Characteristics in Full Scale Horizontal Wellbores	3-21-89 8 members
132	Petroleum Environment Research Forum, No. 87–04 Disposal Alternatives for Spent FCCU Catalysts Uses of spent fluid catalysts	5-1-89 5 members
133	Recording Industry Association of America Copyright protection	5-1-89 11 members
134	Bellcore/PictureTel Corporation Video communication equipment and services	5-10-89 2 members
135	Bellcore/VideoTelcom Corporation Video communication services and equipment	5-10-89 2 members
136	National Center for Advanced Technologies Composite materials, integrated circuits, artificial intelligence	5-30-89 52 members
137	Lehn and Fink Products Group-Aerosol Classification Aerosol products	5-31-89 2 members
138	Bellcore/Bell-Northern Research Moving video	6-15-89 2 members
139	Southwest Research Institute-Development of Computer-Aided Armor Design/Analysis Systems Armor design software	7-2-89 7 members
140	Bellcore/Telecomm Research Laboratory Exchange and exchange access services for optical communications	7-27-89 2 members
141	Bellcore/Samsung Software Exchange access services	8-17-89 2 members

Number	Venture Name/ Research Focus	Federal Registration Number of Members
142	Southwest Research Institute-Forecast Update of the United States Transportation Fuel Quality Forecasts of future fuel quality	9-1-89 4 members
143	American Iron and Steel Institute AISI Direct Steelmaking Project	9-11-89 3 members*
144	Consortium for Superconducting Electronics Superconducting electronics applications	9-18-89 3 members
145	Southwest Research Institute-Development and Demonstration of Seismic Sources for High Resolution Interwell Imaging Borehold seismic instrumentation	9-18-89 9 members
146	The University of Houston/MCC High-temperature superconductivity	9-18-89 9 members
147	Bellcore/Conductus, Inc. Applications for exchange access services	10-17-89 2 members
148	Bellcore/Northern Telecom, Inc. Improvement in exchange access services	10-17-89 2 members
149	Bellcore/AT&T Audio and video teleconferencing equipment and services	10-17-89 2 members
150	Bellcore/AT&T Digital information transmission over subscriber loops	10-17-89 2 members
151	PERF: Project 88–07 Basic Principles and Controls of Refinery Emulsion Formation	10-17-89 10 members
152	Appliance Industry/Government CFC Replacement Consortium CFC11 and 12 replacements	11-1-89 8 members
153	Advanced TV Test Center Advanced television service	11-6-89 8 members
154	Corning/NGK Joint Project New catalyst supports	11-8-89 2 members
155	Advanced Television Test Center/Cable Television Laboratories, Inc. Advanced television service	11-29-89 2 members
156	Automotive Emissions Cooperative Research Program Emission control systems	11-29-89 17 members

Number	Venture Name/ Research Focus	Federal Registration	Number of Members
157	Engine Manufacturers Association/Southwest Research Institute Catalytic converter systems	11-20-89	10 members
158	MRI/Pyrolysis Materials Research Breakdown, processing and /or conversion of chemicals	12-8-89	6 members
159	Advanced Helicopter Electromagnetic Program Analytical methods for advanced helicopter applications	1-10-90	10 members
160	Bellcore/Toshiba Asynchronous transfer mode technology	1-19-90	2 members
161	Bellcore/Furukawa Multi-quantum well lasers	1-19-90	2 members
162	National Food Lab Tamper evident closures	2-12-90	5 members
163	Allethrin Joint Venture Allethrin (pesticide) testing	2-22-90	2 members
164	Bellcore/Siemens Digital video coding and subscriber line concepts	2-22-90	2 members
165	Bellcore/Industrial Technology Research Institute Digital video coding and subscriber line concepts	2-22-90	2 members
166	Power Applications Research Center High-energy electrical cable configurations	2-28-90	2 members
167	Halon Alternatives Research Halon alternatives	3-7-90	6 members
168	Jewelry Manufacturers Project Jewelry manufacturing processes	3-16-90	10 members
169	Advanced TV Research Consortium Development of advanced television systems	3-22-90	4 members
170	Bellcore/Societa Cavi Pirelli, S.A. Optical amplifiers and fiber components	3-26-90	2 members
171	Bellcore/AT&T Low-threshold current surface emitting lasers	3-26-90	2 members
172	Surface Cleaning Technology Consortium Ultra-clean surfaces	3-26-90	3 members
173	International Magnesium Development Corporation Formation and nature of corrosion of magnesium	3-27-90	81 members*

Number	Venture Name/ Research Focus	Federal Registration Number of Members
174	Petroleum Environmental Research Forum: No. 88–06 Study Effect of Disposal of Waste Freshwater Drilling Fluid in Earthen Pits During Operation and After Closure	4-2-90 10 members
175	SQL Access Group Standard set of computer software specifications	4-5-90 17 members
176	1990 Horizontal Well Gravel Pack Program Gravel packing of horizontal oil wells	4-5-90 12 members
177	Petroleum Environmental Research Forum No. 88–09 emergency response manual for natural resource damage assessment	4-18-90 10 members
178	Industry Coop. for Ozone Layer Protection Safe, environmentally acceptable alternatives for ozone depleting substances	4-18-90 14 members
179	C. R. Bard, Inc. E. I. DuPont Improvements of catheters and medical devices	4-23-90 2 members
180	Petroleum Environmental Research Forum No. 88–04 Book re: bioremediation of soils containing petroleum	5-1-90 10 members
181	U.S. Export Council for Renewable Energy Fossil fuel and radical deforestation alternatives	5-30-90 11 members
182	Consortium for Superconducting Electronics Applications of superconducting electronics	6-13-90 3 members
183	Bellcore/Chisso Corporation Electron beam resist materials	6-29-90 2 members
184	Bellcore/Nippon Telegraph and Telephone Telecommunications	6-29-90 2 members
185	Bellcore/Plassey-UK Ltd. Integrated optoelectronic technology	6-29-90 2 members
186	Bellcore/Alcatel N. V. Exchange and exchange access services	6-29-90 2 members
187	Spray Drift Task Force Development of generic spray drift data	7-5-90 8 members
188	Ethanol Joint Venture Toxicological research on ethanol	7-5-90 15 members
189	Petroleum Environmental Research Forum No. 88–05 In Situ Reclamation of Oily Pits	7-19-90 10 members

Number	Venture Name/ Research Focus	Federal Registration	Number of Members
190	NCR Corporation/Teradata Corporation Parallel processing systems	8-20-90	2 members
191	Air Conditioning and Refrigeration Research University of Illinois Air conditioning and refrigeration equipment	8-20-90	15 members
192	Bellcore/Electric Power Research Institute Thin film superconductors	8-22-90	2 members
193	Bellcore/Telecommunications Research Laboratory Broadband telecommunications	8-22-90	2 members
194	Michigan Materials and Processing Institute Polymer-based composites	9-6-90	17 members
195	International Pharmaceutical Aerosol Consortium for Toxicology Testing Testing of HFA-134a	9-6-90	7 members
196	Specialty Metals Processing Consortium Specialty metals processing advancement	9-17-90	10 members
197	Exxon Production Research Company Loop current research	9-18-90	10 members
198	Bellcore/VLSI Application of advanced CMOS VLSI technology to telecommunications	9-18-90	
199	Massachusetts Institute of Technology/TV of Tomorrow Consortium New systems and applications for advanced computer/TV technology	9-27-90	7 members
200	Fuel Cell Commercialization Group Production of electrical energy by molten carbonate fuel cells	10-25-90	11 members
201	Cable TV Labs/General Instrument Corporation Conduct visual degradation tests for cable TV pix	11-1-90	2 members
202	Amoco/ARCO Neural network technology to geophysical exploration	11-7-90	2 members
203	Industrial Macromolecular Crystallography Association Special equipment for x-ray crystallography research	12-3-90	7 members
204	Bellcore/Symbolics, Inc. Advanced animation software systems	12-11-90	2 members

* membership is composed of multimember organizations.

Appendix II

Consortia Filings and Memberships

Summary of Consortia Filings

New consortia filings per year

1985	50
1986	17
1987	25
1988	33
1989	33
1990 (November)	<u>46</u>
Total	204

Consortia Membership

2 members	75	37%
3–5 members	34	17%
6–10 members	45	22%
more than 10 members	<u>50</u>	<u>24%</u>
Total	204	100%

About the Editors

David V. Gibson is associate director, Center for Technology Venturing, College and Graduate School of Business and a research fellow at the IC^2 Institute at the University of Texas at Austin He received his B.A. from Temple University, an M.A. from Pennsylvania State University, and an M.A. from Stanford. In 1983 he earned a Ph.D. from Stanford after completing studies in the areas of organizational behavior and communication theory.

Dr. Gibson is co-director of the Multidisciplinary Technology Transfer Research Group at the University of Texas. He teaches undergraduate and graduate courses on communication behavior in organizations, international business, technology transfer, the management of technology and information systems, and research methods. He belongs to the following professional associations: the Academy of Management, the American Sociological Association, the International Communication Association, and TIMS.ORSA (College on Innovation Management and Entrepreneurship).

Dr. Gibson's research and publications focus on the strategic management of information systems, cross-cultural communication and management, and the management and diffusion of innovation. He has published in the *Journal of Business Communications, Journal of Business Venturing, Journal of Technology Transfer, IEEE Transactions on Engineering Management, Journal of Engineering and Technology Management,* and *Journal of Organizational Computing.* Dr. Gibson is a consultant to business and government and has made professional and keynote presentations in France, Ireland, Italy, England, Egypt, and Taiwan. Dr. Gibson is co-editor of *Creating the Technopolis: Linking Technology Commercialization and Economic Development* (Ballinger, 1988), *Technology Transfer: A Communication Perspective* (Sage, 1990), *University Spin-Off Companies: Economic Development, Faculty Entrepreneurs, and Technology Transfer* (Rowman and Littlefield, 1991) and editor of *Technology Companies and Global Markets: Programs, Policies, and Strategies to Accelerate Innovation and Entrepreneurship* (Rowman and Littlefield, 1991).

Raymond W. Smilor is the executive director of the IC^2 Institute at the University of Texas at Austin. He is also associate professor of management in the UT Graduate School of Business where he directs the Entrepreneurial Studies Program.

Dr. Smilor has published extensively with refereed articles in journals such as *IEEE Transactions on Engineering Management, Journal of Business Venturing,* and *Journal of Technology Transfer.* His research areas include technology transfer, entrepreneurship, economic development, and technology management and marketing. His works have been translated into Japanese, French, and Russian.

He is the author or editor of nine books, including *Corporate Creativity* (Praeger, 1984), *Financing and Managing Fast-Growth Companies: The Venture Capital Process* (Lexington Books, 1985), *The Art and Science of Entrepreneurship* (Ballinger, 1987), *The New Business Incubator* (Lexington Books, 1986), *Creating the Technopolis* (Ballinger, 1988), and *Customer-Driven Marketing: Lessons from Entrepreneurial Technology Companies* (Lexington Books, 1989).

He is a consultant to business and government and has lectured in China, Japan, Canada, England, France, Italy, and Australia. He is adjunct professor of entrepreneurial studies at the School of Business, Bond University in Australia. He has been a leading participant in the planning and organization of many regional, national, and international conferences, symposia, and workshops. He helped to establish and then served as chairman of the College on Innovation Management and Entrepreneurship of the Institute of Management Science.

Dr. Smilor speaks extensively to business, professional, and academic groups in the United States. He is also involved in several civic professional organizations and appears in *Who's Who in the South and Southwest.*

Dr. Smilor has taught management and marketing courses at the UT College and Graduate School of Business. His Entrepreneurship and Venture Competition courses are rated by graduate students among the best in the M.B.A. program. In 1991, he received the Teaching Innovation Award for his development of new teaching methodologies for entrepreneurship. He is also one of the most highly rated instructors in the Management Development Program in the UT Graduate School of Business. He earned his Ph.D. in U.S. history at the University of Texas at Austin.

About the Contributors

J. Grant Brewen is president and chief executive office of the Biotechnology Research and Development Corporation, Peoria, Illinois. He has published over seventy scientific articles and holds three patents. He was educated at Johns Hopkins University, Baltimore, Maryland; B.A. 1961 (honors), Ph.D. 1963, in biology with a major in cell biology.

Bill Curtis is the director of the Software Process Program in the Software Engineering Institute at Carnegie Mellon University. His group is building methods and technology for helping organizations mature their process for building software systems. He was formerly director of the Software Technology Program at MCC. Dr. Curtis is an associate editor of *IEEE Software, Human-Computer Interaction, International Journal of Man-Machine Studies,* and *Journal of Systems and Software*.

Claude DelFosse is vice president of technology transfer for the Software Productivity Consortium. He joined the Consortium in September 1986, as a principal member of the Software Product Transfer Division; he was named vice president of technology transfer in June 1989. Mr. DelFosse holds advanced degrees in business and aerospace engineering from the University of Paris and the California Institute of Technology, respectively. He has been recognized for outstanding scholarship with both Fulbright and NATO fellowships.

Grant A. Dove is chairman of the board of the Microelectronics and Computer Corporation Technology Corporation (MCC). He also served as chief executive officer of MCC from 1987 to 1991. During his tenure at MCC, Mr. Dove has overseen the addition of start-up programs in High Temperature Superconductivity and Optics in Computing. He was closely involved with replanning MCC's Computer-Aided Design Program and participated in restructuring of the Advanced Computer Architecture and Packaging/Interconnect Programs. Mr. Dove received his bachelor of science in electrical engineering from Virginia Polytechnic Institute in 1951. He is a member of Eta Kappa Nu, Tau Beta Pi, and Phi Kappa Phi.

Alan Engel has been the overseas representative of ERATO, the Research Development Corporation of Japan, since 1986. As president of International Science and Technology Associates, Inc. (ISTA), Dr. Engel devotes his time to international science and technology strategy, negotiations, and information centering on Japan. Dr. Engel holds a Ph.D. in physical biochemistry from the Rockefeller University.

William M. Evan is professor of sociology and management at the University of Pennsylvania. His principal current research and teaching interests are organization theory, sociology of law, and problems of war and peace. He received his B.A. at the University of Pennsylvania and his Ph.D. at Cornell University.

Wendell M. Fields is the manager of the Software Initiative Technology Transition Program in Hewlett-Packard's Corporate Engineering. This group is responsible for designing and implementing a strategy to transfer methods, tools, development environments, consulting, metrics, and training in focused areas of software engineering. He received his bachelors degree from Brigham Young University, his Masters from the College of Notre Dame, and his doctorate from the University of San Francisco in the areas of organizational behavior and educational psychology.

Hervé Gallaire is the vice president of software development, GSI in Paris, France. He received his Ph.D. from the University of California at Berkeley in 1968.

John F. Hesselberth is vice president, Du Pont Fibers Research and Development. His experience covers a wide range of technical and process technology assignments relating to all of the Fibers businesses. Dr. Hesselberth is a chemical engineer with degrees from Purdue University and the University of Cincinnati.

Louis D. Higgs is president of L.D. Higgs & Associates in Albuquerque, New Mexico and is a senior fellow, Technology and International Trade, for the Center for the New West in Denver, Colorado. He has over twenty-five years of professional experience in R&D and interinstitutional arrangements, has served both public and private institutions including federal, regional and state agencies, universities, and major international companies, as well as working with a number of technology oriented new business start-ups. Mr. Higgs has graduate degrees in the philosophy of science and international relations.

Fumihiko Kamijo is a professor at the Information Science Laboratory, Tokai University in Japan. He is also a member of the Information Processing Policy Committee.

Takashi Kurozumi is deputy director of the Research Center, ICOT, in Japan. Mr. Kurozumi holds a master degree in electro-engineering from Kyoto University.

Mark S. Lieberman is the associate deputy secretary of commerce at the U.S. Department of Commerce. Mr. Lieberman's focus is on technology, space, and telecommunications, helping to manage agencies within the Department of Commerce to ensure that policies remain consistent with the Secretary's goals. Mr.

Lieberman graduated magna cum laude with a bachelor of science in mechanical engineering from Tufts University, College of Engineering. He later graduated from the Benjamin N. Cardozo School of Law in New York City.

Edward A. Miller is president, National Center for Manufacturing Sciences. His 24-year career includes experience in corporate management and manufacturing technology development. Mr. Miller received his B.S. in mechanical engineering from Western New England College.

Larry Novak is external technology manager, MOS Memory Products, Semiconductor Group, Texas Instruments. He has worldwide responsibility to coordinate the equipment improvement activities, definition of equipment-related methodologies, supplier interfaces, and commonality among semiconductor manufacturing facilities. He spent February 1988 to February 1991 as an assignee to SEMATECH from Texas Instruments. He earned a B.S. in physics in 1966 and an M.S. in engineering management in 1973 from the University of Tulsa. Mr. Novak holds a patent for a TI Calculator IC.

Robert Noyce was founding president and chief executive officer of SEMATECH from 1989 to 1990. Dr. Noyce was a co-inventor of the integrated circuit and was co-founder of Intel Corporation. He was a founding member and first president of the Semiconductor Industry Association. He earned his Ph.D. in physical electronics at the Massachusetts Institute of Technology. Dr. Noyce died June 9, 1990 and is remembered as an inventive engineer, visionary entrepreneur, concerned American, and one of the fathers of Silicon Valley.

Paul Olk is an assistant professor at the graduate School of Management of the University of California–Irvine. In addition to continuing the examination the consortium formation and management, his research activities include studies on organization creation, U.S. competitiveness, and top executive succession. He received his doctorate in the area of organization and strategy from the Wharton School of Business of the University of Pennsylvania.

Debra M. Amidon Rogers is a senior management systems research program manager at Digital Equipment Corporation. In this capacity, she enables a worldwide transformation strategy with modern enterprise innovation technology. She has published numerous articles, reference manuals, and technical reports. She holds degrees from Boston University, Columbia University, and MIT, and received several awards including Pi Lambda Theta Scholar, "The Outstanding Young Professional of New England," and the President's Award from the Technology Transfer Society.

Michael S. Rubin is president of Molinaro/Rubin Associates, an international management consulting firm specializing in partnership strategies and project development. He is currently advising development consortia and partnerships in the United States, France, and Korea. Since 1983, Mr. Rubin has been on the faculty at the University of Pennsylvania's Fels School of Government. He received his Ph.D. from the Wharton School of Business, the University of Pennsylvania.

Dieter Schütt is director, Systems and Networks, Corporate R&D of Siemens AG, Munich, Federal Republic of Germany, and an adjunct professor of the faculty for computer science at the University of Erlangen–Nurnberg. Dr. Schütt received his Ph.D. in mathematics in 1970 and Dr. Habil in computer science in 1983 at the University of Bonn.

Syed Z. Shariq is assistant director for Science and Industry at NASA-Ames Research Center. He is currently on assignment as the chief executive officer of American Technology Initiative, a nonprofit corporation dedicated to the implementation of R&D joint ventures between the public and private sectors. Dr. Shariq holds a Ph.D. in operations research from Virginia Polytechnical Institute and State University.

Warren D. Siemens is deputy to the vice president of technology transfer at Martin Marietta Energy Systems, Inc., in Oak Ridge, Tennessee. Dr. Siemens holds B.S. and Ph.D. degrees from the Massachusetts Institute of Technology.

About the Sponsors

The IC² Institute is a major research center for the study of innovation, creativity, and capital, hence IC². The institute studies and analyzes information about the market system through an integrated program of research, conferences, and publications.

The key areas of research and study concentration of IC² include the management of technology; creative and innovative management; measuring the state of society; dynamic business development and entrepreneurship; econometrics, economic analysis, and management sciences; and the evaluation of attitudes, opinions, and concerns on key issues.

The institute generates a strong interaction between scholarly development and real-world issues by conducting national and international conferences, developing initiatives for private- and public-sector consideration, assisting in the establishment of professional organizations and other research institutes and centers, and maintaining collaborative efforts with universities, communities, states, and government agencies.

IC² research is published through monographs, policy papers, technical working papers, research articles, and four major book series.

The College and Graduate School of Business prepares outstanding students for a variety of careers and early assumption of management responsibilities in a rapidly changing environment. A continually evolving curriculum ensures dynamic programs that meet the needs of future business leaders. A broad range of elective opportunities enables the student to take a generalist approach to advanced study, to concentrate in a traditional or emerging field of business study, or to broaden perspectives through supporting course work in nonbusiness disciplines.

The College of Business Administration and Graduate School of Business were created in 1922. The college has five departments: accounting, finance, management science and information systems, management, and marketing administration. Approximately 8,736 undergraduate students and 1,505 graduate students are currently enrolled in the College and Graduate School of Business.

The College of Communication was created almost twenty-five years ago and includes the departments of advertising, journalism, radio-television-film, and speech communication, as well as the Center for Research on Communication Technology and Society and the Center for Telecommunication Services (CTS), which houses National Public Radio members station KUT-FM, the Longhorn Radio Network, and the UT cable system.

The college recognizes the inevitable impact that new technologies will have on our society in the future, especially on the field of communication. Consequently, the college is committed to the education of students and to research that takes these technological changes into consideration and will produce students who will assume leadership roles in various fields of communication and in the nation's institutions of higher education.

The fall 1988 enrollment totaled 4,162 students—3,659 undergraduates and 503 graduate students—supported by over 80 faculty members. The College of Communication currently has six endowed chairs, twenty endowed professorships, and twelve endowed teaching fellowships and lectureships.

The Microelectronics and Computer Technology Corporation (MCC) is a cooperative enterprise whose mission is to strengthen and sustain America's competitiveness in information technologies. Its objective is excellence in meeting broad industry needs through application-driven research and development and timely deployment of innovative technology. MCC has nineteen shareholders, or equity participants, seventeen associate members, and two government sponsors. MCC conducts research in five areas: advanced computing technology, computer-aided design, electronic applications of high-temperature superconductors, semiconductor packaging/interconnect, and software technology.

The National Center for Manufacturing Sciences (NCMS) is a consortium of more than ninety leading corporations committed to making U.S. manufacturing globally competitive again. It is a cooperative effort marshaling the support for some of the nation's leading manufacturers, its legislators, its educators, and its businesspeople. Through these partnerships, NCMS funds research projects to develop the next generation of manufacturing technology.

Member companies and NCMS staff direct these projects within a strategic framework. NCMS then locates the best facilities to carry out and coordinate the research efforts. The results are then made available to member companies. NCMS plays an important role in technology transfer by further promoting the implication of research findings among member companies and their suppliers and customers.

KPMG Peat Marwick is a member of KPMG, a major worldwide professional services organization, with more than 63,000 employees and 650 offices operating in some 120 countries. The firm provides auditing and accounting, tax, and management consulting services. KPMG Peat Marwick's sponsorship of this conference is through its technology development services practice area—a collaborative program of its specialized industry practices in high technology, higher education, and government services.

KPMG Peat Marwick established its special consulting initiative in Technology Development Services (TDS) to provide leadership in a wide range of issues pertaining to technology, innovation, and competitiveness at the national, state, and local levels. TDS services are focused on universities and colleges, research institutions and federal labs, government agencies, and other entities involved with research and technology development, transfer, management, and commercialization.

KPMG Peat Marwick's High-Technology Practice provides specialized services to high-technology companies of all sizes and at all stages of development. The firm's Higher Education Practice serves over 600 university and college clients nationwide. KPMG Peat Marwick's Government Services specialists advise over 2,300 clients at the state and local levels in the development of a wide range of technical tools and systems required for the efficient and effective management of public programs.

The Technology Transfer Society was formed in 1975. It is a nonprofit international organization created to support the effective forecasting, transfer, utilization, and assessment of technology. The society serves a multidisciplinary membership of more than 600 professionals, dedicated to the efficient movement of technology for the purposes of economic growth and an improved quality of life for all people. Through its peer-review journal, its newsletter and selected publications, its workshops, and annual symposiums the society provides a dynamic forum for new ideas in the technology transfer area.

The Center for the New West is an independent, nonprofit research institution established in Denver in January 1989. The work of the center focuses on public policy and strategies for economic development, especially as they affect the western region of the United States.

The center emphasizes opportunity for small and medium-sized business in the "new economy." (The term "new economy" refers to the ongoing restructuring that increasingly characterizes our society—restructuring that is the result of expanding global competition, dramatic demographic shifts, technological change, rapidly changing consumer tastes, entrepreneurship, and the growing impact of innovation on enterprise formation.) The center seeks to improve the quality and usefulness of information about the "new economy" as well as help businesses and government leaders in nineteen western states, from the Mississippi River to Hawaii, be more responsible to these changes in our society.

The center is supported by US West, Inc., El Pomar Foundation, *The Denver Post,* Public Service Company of New Mexico, Salt River Project, MIC Telecommunications Corporation, Goldman, Sachs and Company, Coopers & Lybrand, and other companies in cooperation with public-sector authorities including the Western Governors' Association, and the Western U.S. Senate Coalition.

The RGK Foundation was established in 1966 to provide support for medical and educational research. Major emphasis has been placed on the research of connective tissue diseases, particularly scleroderma. The foundation also supports workshops and conferences at educational institutions through which the role of business in American society is examined. Such conferences have been cosponsored with major universities and research institutions in the United States and in other countries.

The RGK Foundation Building has a research library and provides research space for scholars in residence. The building's extensive conference facilities have been used to conduct national and international conferences. Conferences at the RGK Foundation are designed to enhance information exchange on particular topics and to maintain an interlinkage among business, academia, communities, and government.